William J. (William John) Fryer

Architectural Iron Work

William J. (William John) Fryer

Architectural Iron Work

ISBN/EAN: 9783743687059

Printed in Europe, USA, Canada, Australia, Japan

Cover: Foto ©berggeist007 / pixelio.de

More available books at **www.hansebooks.com**

ARCHITECTURAL IRON WORK.

A PRACTICAL WORK

FOR

IRON WORKERS,

ARCHITECTS, AND ENGINEERS,

AND ALL WHOSE TRADE, PROFESSION, OR BUSINESS CONNECTS
THEM WITH ARCHITECTURAL IRON WORK.

SHOWING

THE ORGANIZATION AND MECHANICAL AND FINANCIAL
MANAGEMENT OF A FOUNDRY AND SHOPS

FOR THE MANUFACTURE OF

IRON WORK FOR BUILDINGS,

WITH

SPECIFICATIONS OF IRON WORK,

USEFUL TABLES,

AND

VALUABLE SUGGESTIONS FOR THE SUCCESSFUL CONDUCT
OF THE BUSINESS.

By WM. J. FRYER, Jr.

NEW YORK:
JOHN WILEY & SONS.
1876.

234×2

DEDICATED

<table>
<tr><td>TO</td><td></td><td>TO</td></tr>
<tr><td></td><td>AND</td><td></td></tr>
<tr><td>*CAPITAL,*</td><td></td><td>*LABOR,*</td></tr>
</table>

IN THE PERSON OF	IN THE PERSONS OF
Eugene Kelly, Esq.,	My former fellow Workmen,
BANKER,	WHO EXTENDED TO ME THEIR SYM-
A CHRISTIAN GENTLEMAN, A TRUE	PATHY IN THE MOST TRYING
FRIEND, A WISE AND SAGACIOUS	PERIOD OF MY LIFE.
COUNSELOR.	

THE AUTHOR.

CONTENTS.

vi CONTENTS.

CONTENTS. vii

ARCHITECTURAL IRON WORK.

GENERAL REMARKS.

ESTABLISHMENTS devoted exclusively to the manufacture of iron work for buildings are of comparatively recent growth. Almost without exception the larger ones now existing have grown from small beginnings, building after building having been added to the original shop until they became great workshops without proper plan for the economical working and handling of materials. Formerly there were two distinct divisions in contracting the iron work required for a building; the wrought iron was given to a blacksmith, and the cast iron work to a foundryman. The custom now is to give the entire work to one establishment.

This branch of iron manufacture has increased enormously within the past fifteen years, and the probabilities are that the future will develop a still greater proportional growth. It is a commonplace saying that as a nation we have but just begun to use iron. This is, indeed, very true as regards its use for building purposes. Good construction, economy of material, and beauty of form in architectural iron work have made greater progress in this country, and particularly in the city of New York, than elsewhere in the world. A knowledge of the subject requires diffusion. Years of study, observation, and hard practical toil were the price of the author's thorough knowledge of this class of work, as it must be to every man who would qualify himself for this business.

The aim in giving publicity to this knowledge is largely for the enlightenment and advancement of workingmen. They need to have placed before them in plain and intelligible forms

1

an outline of how the works in which they daily toil are managed, and so to help educate up operative mechanics to become competent to command and control the coming great industrial workshops of our land.

To proprietors of works new light will be thrown on their business, and enable them more thoroughly to understand the principles which govern their every-day doings. The knowledge herein imparted will enable a manufacturer to correctly ascertain what his products cost, and to establish prices which will allow fair profits. It is a general complaint that the cost of work almost invariably exceeds an estimate, and the yearly balance-sheets too often indicate that a business has failed to pay a reasonable reward for the labor and use of capital employed. The cost of the various items given in the following pages will differ more or less in every establishment; but if the principles laid down will induce manufacturers of iron work for buildings to make similar statements of actual costs, in detail, applicable to their own shops, there will be little danger that their products will be sold without profit, or that the balance at the end of the year will be found on the wrong side.

A MODEL SHOP.

LOCATION.

In selecting a site for the shops many essential things are to be considered. The land should have a water frontage on a navigable stream, be convenient to railroad depots and steamboat landings, have good telegraphic and mail connections, and be where skilled labor is easily obtained, and where homes for workingmen are numerous. A good-sized plot of ground is desirable, not alone for the immediate present, but to accommodate the future growth and requirements of the business. The land must be of moderate value, and selected with an eye to its prospective increase in value. Look ahead to a profit on the land purchase. It is well to have the location

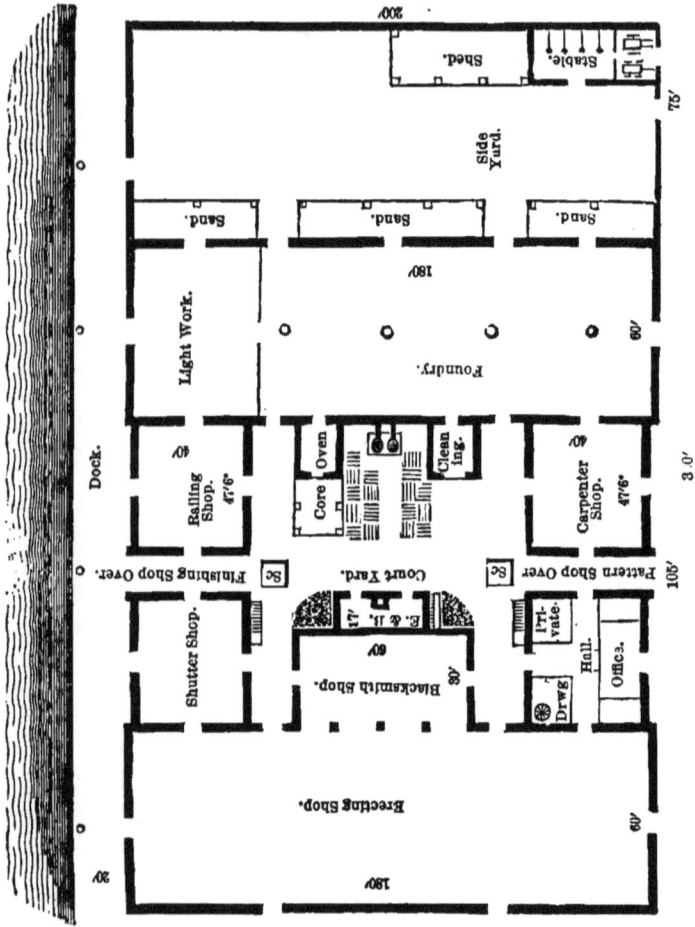

Height of Stories in Clear.

Foundry, 22'.   Blacksmith Shop, 20'.   Pattern Shop, 10'.   Finishing Shop, 10'.

Erecting Shop, 22'   Office, 11'.   Railing Shop, 11'.   Engine Room, 16'.

Labels appearing on plan: Dock. — Light Work. — Foundry. — Railing Shop. — Core Oven — Cleaning. — Carpenter Shop. — Finishing Shop Over. — Court Yard. — Pattern Shop Over — Shutter Shop. — Blacksmith Shop. — Private. — Hall. — Drwg. — Office. — Erecting Shop. — Sand. — Side Yard. — Shed. — Stable. — F. & B. — Sc.

Dimensions: 200' — 180' — 105' — 60' — 40' — 47'6" — 30' — 3.0' — 75' — 17'.

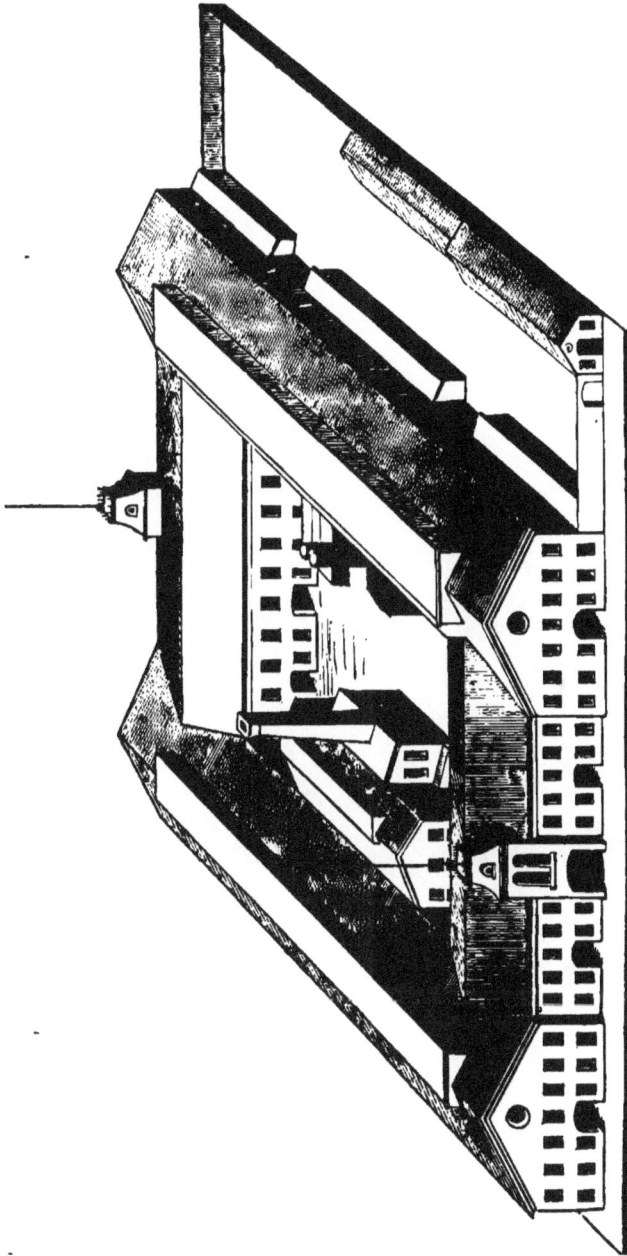

PERSPECTIVE VIEW AND PLAN OF A MODEL FOUNDRY AND SHOPS FOR ARCHITECTURAL IRON WORK.

away from other shops in the same line of manufacture, so as to draw employés to the neighborhood and secure their permanency; and yet be not so far away as to greatly inconvenience temporary hands.

Selecting such a plot of ground, of a size not less than 300 × 250 feet, suppose its cost to be $15,000.

BUILDINGS.

The buildings will all be of brick, with double-pitch frame roofs, covered with slate, and put up in a good and substantial manner. Their cost may be taken at $40,000.

The shops are arranged in relation to each other as to insure the least handling and inconvenience from the time the raw material is landed on the dock until the manufactured article is run out for shipment—one succession of advances. By reference to the plan it will be seen that the buildings form a hollow square. This secures the greatest amount of light and ventilation, the greatest security to valuable materials, the least danger from destructive fire, and the best control of the employés.

The engine and boiler are situated at the centre, the power radiating to all quarters. The cupolas are placed at the centre of the length of the foundry, and the run way for charging the same is in the yard. The foundry is 60 × 180 feet. A portion of it, fifty feet in length, is railed off for light work. The remainder, for heavy work, is furnished with four cranes. Sheds for sand run alongside the foundry, and the sand is thrown directly in as required. Two cupolas are provided, each with a maximum capacity of twenty tons, enabling a cast to be made every working day in the year without having to lay by during relining, etc. By using both at once sufficient iron can be melted for almost any purpose. The erecting shop, in which to lay down iron fronts and other work is 60 × 180 feet. The blacksmith shop is 30 × 60 feet, and opens out into the erecting shop, so that the latter may be used for purposes

connected with the former. A finishing shop is made two
stories in height, in which to make shutters, railings, and fit
up small work. The second story floors will be suspended by
iron rods from the roof trusses, so that the first story shall be
entirely free from columns. The opposite building is also
made two stories in height, a portion of its first floor being
used for a carpenter and a flask-making shop, and the upper
story, 40 × 105 feet, for pattern making. The stairs thereto is
on the outside of the building. On the first floor of this building is arranged the offices ; a main counting-room, a private
office, and a drawing-room, the latter connecting with the
pattern shop by a circular stairway. From the windows of
the private office a general survey of the premises is obtained.
Drive-ways through the shops are plentifully provided, and
weighing scales are so placed as to accommodate incoming and
outgoing materials, and for the weighing of rough castings in
transit from the foundry to the finishing shop. The core oven,
14 × 20 feet, is placed near the cupolas, together with a house
for core-making. On the other side of the cupolas is a small
house for brushing and cleaning castings. It will be advisable
in localities where winters are severe to roof over the yard or
court, taking care to provide as much light as possible and
liberal ventilation. The roof can then be made use of for
storage of small flasks and similar things. A stable, 20 × 50
feet, is placed where shown in the side yard. This yard gives
space for flasks, cord-wood, etc. The workmen in going out
and in daily all pass through the entrance way alongside of
the office. When being paid off they pass through the hall-
way and main office.

Enlarged capacity to the shops can be had by adding a
wing on the foundry, covering more or less of the side yard.
The erecting shop may have a gallery added, fifteen feet wide,
running around on all sides, suspended from the roof trusses,
and used for vice work. In due time a two or three story

building for storage of patterns will be required; this will be built on a portion of the space of the side yard, and will be disconnected from the other buildings.

An iron-works planned as shown and described would, for its purpose, be superior to any existing at the present time, and its capacity, in proportion to its cost, be far ahead of any.

FINANCIAL MANAGEMENT.

The iron business is a heavy business, and to manufacture in a first-class way requires a large capital. Whatever amount of money is put into the venture—and it is a venture, as all business operations are—be it remembered that this capital is worth seven per cent. per annum, for that interest can be obtained without risk and without trouble. Then there are expenses connected which are inevitable and constant, whether much or little is done. Taxes, insurance, office employés, expenses of running engine, pay to foremen, etc.; these go on about the same whether 100 or 300 men are employed as producers—the same on $100,000 as on $300,000 worth of work. Above a certain limit on a given investment, the difference between the cost of the raw materials and labor employed, and the prices obtained for the finished articles, is the profit. Therefore, one of the secrets of making money is to keep the works filled to their utmost capacity.

To illustrate this principle, suppose that a lot of columns, twelve inches in diameter and three-quarters of an inch in thickness, are to be made at a given price—say, four cents per pound. Now, if these same columns were to be made one-and-a-half inches thick, and the rate per pound was the same, the heavier weights would afford by far the best profit, because the cost in both cases are alike as to moulding time, and materials, cleaning, chipping, turning off ends, etc., and the heavier weights represent simply melted pig iron poured into the mould. There is danger, however, of these facts leading a contractor astray, and tempting him to take work too low. A limit must, therefore, be established; and when a man is steeled to refuse work below that limit, and yet has the energy and ability to keep the shops well filled with contract work above that limit, good results may confidently be looked for at the end of the fiscal year. If a job be taken at an unprofitable figure, no amount of drive can overcome the error, whether intentional or unintentional, made at the start. But whether a job is taken at a good price or a poor one, never slight the work. Always do the best that can be done, both in material and in execution. A reputation for good castings and true fitting will, in due time, become extensively known, and turn the scales of owners' preference in giving such an establishment work where estimates run close. The expense of doing good work is no greater, and perhaps not as great, as to do botch work. If the workmen are held up to a proper standard, and whenever a mechanic shows himself incompetent or careless, he be discharged and replaced with a better man, the entire force will do their work in a thorough and expeditious manner. If any journeyman be addicted to drink, no matter how good a mechanic he may be, or if he is disputative or loud in his political preferences or religious views, it is well to weed out all such and be free of them.

The cost of ground and buildings has been set down as $55,000. The machinery will require an expenditure of $45,000, and a working capital, over and above all, of $50,000. Thus the establishment is supposed to represent $150,000. Expenses will commence with the organization, and go on unceasingly. These are to be taken into account and apportioned to the different shops. They become what will be termed shop expenses—so much on the foundry, so much on the finishing shop, etc., in proportion to the room they occupy.

The cost of castings in the foundry wants to be got at. To one unfamiliar with a foundry—perhaps to many familiar with a foundry—this would appear a very difficult task. And yet for a given month, if a record of the quantity and cost of the pig iron consumed, together with the sand, flour, wood, coal, and other supplies used, and the wages paid to moulders, helpers, etc., be aggregated, and to this sum the shop expenses, before referred to, be added, and the total in dollars and cents be divided by the number of pounds of good castings weighed up coming out of the foundry during the month, it will give, beyond the shadow of a doubt, the average cost per pound of those castings. A little good judgment will separate those castings into three grades—heavy, medium, and light—and the prices to correspond. So simple is the method when systematically pursued. A monthly record so kept will give the average daily consumption of materials and cost of labor to the ton of iron melted. It is also necessary to get at the exact cost per pound of any particular casting, large or small, and the method of doing this will be shown further on.

The same manner of record applied to finishers engaged in fitting up the castings will establish correctly the average cost per pound of finishing certain grades of castings.

In the blacksmith shop a record kept of the coal used, the wages paid, and the wrought iron cut up, will give the average cost per pound for forgings and smiths' work.

In the pattern shop the average cost of each man is obtained through this same principle. For certain classes of finished castings experience will determine the average cost per pound or per ton for pattern work, including pattern materials, such as lumber, hardware, etc.

Suppose an iron front to have been manufactured in the shop and set up at the building and finished complete. The cost has been kept at every stage, and it must now show all this: The total weight; the weight of the heavy castings, such as the columns and the pilasters; the weights of the light castings, such as the arches, cornices, sills, etc.; the cost of the castings as they came out of the foundry; their cost per pound of finishing in the shop; and the cost per pound of setting up and finishing at the building; the cost of painting; the total cost per pound and the total cost in dollars for the front; also the cost per lineal foot and per square foot superficial.

With records like these there is little room left for guess-work. The lack of them accounts for the wide difference in bids from contractors, and affords an explanation for the disappointing results obtained at the end of a year's business on finding little or no profits made or actual losses incurred. Many concerns take work at losing prices through sheer ignorance of what the actual cost is. Every article in the business, and each particular contract, should be reduced in detail to its cost per pound, or per superficial foot, or both. Certain classes of work cost more for the finishing labor than the castings themselves cost. What would seem to be a large price per pound would not give back the manufacturer his money. A contract job may show a loss, or particular parts of it a loss. But future similar mistakes are thus guarded against. Be governed by facts, results actually obtained, and never be influenced by what a competitor takes work at, other than to impel a closer scrutiny into the correctness of the cost or a more economical manner of doing such work. Sooner or later those who defy the teachings of figures, as well as the teachings of experience, will come to grief.

A man goes into this business for the dollars and cents profit which is in it, not for glory. It is a noble business, and affords scope for the best talents—the astuteness of the lawyer, the sound judgment of the merchant, the genius of the mechanician, and the generalship of the soldier. Fame, however, is but incidental to the business, and surely it will not attach to him who fails to make a financial success of his work. In the eagerness and anxiety to secure contracts, and the liability of mistaking or omitting items, the tendency is to figure too low.

Rather do without work than have it at a loss. Let energy and constant attention to business be the levers which secure to an establishment its complement of work at good prices. The argument that work had better be taken at cost than not at all, will do for the indolent man, or the man who has outlived his energy. An iron works requires to be kept constantly going, or it becomes self-consuming. Work at cost pays the interest and taxes and office hire, and keeps the men together and the tools from rusting, and the establishment generally from running behind. But when the manager cannot find sufficient work at remunerative prices, the establishment is too large for that man, or the man too small and incompetent for the establishment. This is a growing country, and foundries can hardly keep pace with the demand for iron work for buildings. The live man can always find work, even in dull times, during panics and wars. It is of the first importance to get remunerative prices.

A good credit would attach itself to an establishment paid for and provided with a working capital as stated; indeed, an almost unlimited credit, if the manager be known as a competent and reliable man.

The sum invested is a large one, and is represented and used as follows:

CAPITAL $150,000.

Ground cost	$15,000	Interest on capital	$10,500
Buildings cost	40,000	Taxes	1,500
Machinery	45,000	Insurance	800
Working capital	50,000	Gas	700
	———	Repairs to buildings	500
	$150,000	Incidentals	1,000
			———
			$15,000

OFFICE EXPENSES.

Wages—Manager	$5,000
Book-keeper	1,500
Time-keeper	800

```
Amount brought forward..............$7,300  ——— $15,000
Two boys ............................   600
Draughtsman .........................  1,200
Night watchman ......................   900
Incidentals..........................  1,000
                                      ———— 11,000
```

ENGINE AND BOILER EXPENSES.

```
Coal, per day ...................................$4 00
Oil, tallow, waste, etc..........................  1 00
Repairs, etc.....................................  2 00
Wages of engineer................................  3 00
Incidentals......................................  1 00
                                                 ———
                                                 $11 00
              is, per annum (300 days),   3,300
                                                ———— 14,300
                                                     ————
Expenses...... ...............................................$29,300
```

APPORTIONED AS FOLLOWS:

Foundry—7-16 of $29,300 is $12,818.75 per annum, or per day....... $42 72
Erecting and finishing shop—7-16 of $29,300 is $12,818.75 per annum,
   or per day.. 42 73
Blacksmith—1-16 of $29,300 is $1,831.25 per annum, or per day...... 6 10
Pattern—1-16 of $29,300 is $1,831.25 per annum, or per day......... 6 11

FOUNDRY.

```
Shop expenses, as stated above, per day...................... .........$42 72
Coal for cupola, 1¼ tons at $6.....................................  9 00
Common sand..........  ...........................................  2 00
White sand.........................................................  0 75
Sea-coal, fire-clay, etc ..........................................  2 00
Flour ....................... ........ .............................  2 00
Repairs to ladles, cupolas, etc....................................  2 50
Wood and coal for core oven .......................................  2 00
Flasks, material in and wages making ..............................  10 00
Wages—1 foreman........................................$6 00
      1 melter.........................................  4 50
      2 helpers, $1.75.................................  3 50
      10 moulders, $3.50  ┐
      25    "       3.00  ├ say.........................133 00
      15 helpers,   1.50  ┘
                                                      ———— 147 00
                                                           ————
                                                           219 97
Iron—Per ton.............................$30 00
     Interest, four months................  0 70
     Lighterage .................. ........  1 00
     Cartage, handling, and short weights...  1 00
                                          ———— 32 70 × 8 tons,  261 60
                                                              ————
                                                              $481 57
```

Amount brought forward..................................... $481 57
Cartages, etc 5 00
Contingencies.................................... .... 3 00

Cost of melting 8 tons, with iron included..................... $480 57

Or. per ton... $61 20

A gross ton of iron (2,240 lbs.) yields 2,000 lbs. in castings ; the rest is wastage, and sprues, gates, etc., which makes the cost, without moulding :

Melting. per ton $28.50, or per lb............................... c. 1.425
Iron,   "   32.70,   "   .... c. 1.635

Cost per lb... c. 3.060

If ten tons is melted, then the cost will be :

Melting per ton, $22.80, or per pound .... c. 1.140
Iron   "   32.70, "   "   .... c. 1.635

Cost per pound without moulding .... c. 2.775

It will thus be seen that the cost of melting proportionately decreases as the amount of iron increases. The heavier the castings the cheaper they can be made. To melt ten tons requires scarcely any additional expense over melting eight tons, with the exception of a little coal.

The shop expenses of the foundry are covered when a certain amount of iron is being melted. But the business becomes profitable only when a greater amount is being melted. In manufacturing iron work for buildings, there are very few articles that can be made up into stock or made in advance. Most of the work is taken under contract, and the different parts made just before they are required at the building. Payments are made by the owners of buildings for whom the work is for, as the work progresses. The money turns very quickly, pay for finished work being usually got before the pig iron of which it is made has to be paid for, if bought on the usual four months' credit. Few bad debts are incurred where the contracts are direct with owners, as new buildings for mercantile purposes are rarely built except as investments or to supply the prosperous demands of commercial firms. A manager can, therefore, make close and safe calculations in arranging his finances. There is no heavy stock of articles to be carried waiting for purchasers.

It will be observed that in the foregoing calculation made for the cost of melted iron, the price of pig iron is taken at $30 per ton. This price is merely taken as an illustration, for the object of these tables and the views given are to offer correct principles and a guide in making up tables and costs applicable to any particular iron works.

The mixtures of iron will vary greatly according to location, availability of certain brands of iron, and foundrymen's ideas and experience. The following mixtures are not given with any great degree of confidence as the best, but simply what has been found to work well in practice.

MIXTURES OF IRON.

```
For  heavy  work—No. 1 American Iron........................ 3 parts.
                  No. 2    "        "    ........................ 2  "
                  Scotch Iron................................ 1 part.

For medium work—No. 1 American Iron........................ 1  "
                  No. 2   "        "    ........................ 1  "
                  Scotch Iron ............................... 1  "

For  small  work—No. 1 American Iron .... .................... 1  "
                  Scotch Iron................................ 2 parts.
```

COAL.

Amount to be used will differ in accordance with hardness and kind—say one ton anthracite coal to six tons of pig iron.

TABLE.

Cost of melted iron with foundry expenses added on same :

Iron at $25 per ton.

```
1 ton iron (2,240 lbs.)..................................................$25 00
4 months' interest.......................................................  0 58
Lighterage..............................................................  1 00
Cartage, handling, and short weight..................................  1 00
                                                                      ───────
                                                                      $27 58
```

Yields 2,000 lbs. good castings ; the rest is wastage, etc.

Iron costs per pound .... c. 1.38
Melting (as obtained under the head of "Foundry") per pound ...... c. 1.64
_____
Cost per pound without moulding expenses.................... c. 3.02

COST OF MELTED IRON.

The following table shows the cost of melted iron, with pig from $20 to $50 per ton, including foundry expenses :

PER TON.					PER LB.		
Cost of 1 tg, per Ton.	Four mos. Interest.	Lighterage.	Cartage, Handling and Short Weight.	Total.	Iron.	Melting.	Total.
$20 00	$0 47	$1 00	$1 00	$22 47	c. 1 12	c. 1.64	c. 2.76
21 00	49	1 00	1 00	23 49	1.17	1.64	2.81
22 00	51	1 00	1 00	24 51	1.23	1.64	2.87
23 00	54	1 00	1 00	25 54	1.28	1.64	2 92
24 00	56	1 00	1 00	26 56	1.33	1.64	2.97
25 00	58	1 00	1 00	27 58	1.38	1.64	3.02
26 00	61	1 00	1 00	28 61	1.43	1.64	3.07
27 00	63	1 00	1 00	29 63	1 48	1.64	3.12
28 00	66	1 00	1 00	30 66	1.53	1.64	3.17
29 00	68	1 00	1 00	31 68	1.58	1.64	3.22
30 00	70	1 00	1 00	32 70	1.63	1.64	3.27
31 00	73	1 00	1 00	33 73	1.69	1.64	3.33
32 00	75	1 00	1 00	34 75	1.74	1.64	3.38
33 00	77	1 00	1 00	35 77	1.79	1.64	3.43
34 00	79	1 00	1 00	36 79	1 84	1.64	3.48
35 00	82	1 00	1 00	37 82	1.80	1.64	3.53
36 00	84	1 00	1 00	38 84	1.94	1.64	3.58
37 00	86	1 00	1 00	39 86	1 99	1.64	3.63
38 00	89	1 00	1 00	40 89	2.05	1.64	3.69
39 00	91	1 00	1 00	41 91	2.10	1.64	3.74
40 00	93	1 00	1 00	42 93	2.15	1.64	3.79
41 00	96	1 00	1 00	43 96	2 20	1.64	3.84
42 00	98	1 00	1 00	44 98	2.25	1.64	3.89
43 00	1 00	1 00	1 00	46 00	2.30	1.64	3 94
44 00	1 03	1 00	1 00	47 03	2.35	1.64	3.99
45 00	1 05	1 00	1 00	48 05	2.40	1.64	4.04
46 00	1 07	1 00	1 00	49 07	2.45	1.64	4 09
47 00	1 10	1 00	1 00	50 10	2.50	1.64	4.14
48 00	1 12	1 00	1 00	51 12	2.55	1.64	4.19
49 00	1 14	1 00	1 00	52 14	2.60	1.64	4.24
50 00	1 17	1 00	1 00	53 17	2.65	1.64	4.29

COST OF CASTINGS.

When cost of moulding does not exceed 1 cent per pound.

Moulding........	c. 1.00
Facings, cores, and chaplets.......................................	.20
Cleaning and chipping...	.20
Labor and handling...	.05
Cartage ..:..	.10
Losage on bad castings (10 per cent. of above items)................	.15
	c. 1.70

Melted iron, with shop expenses added (as obtained under head of
" Foundry "). Pig iron calculated on a basis of $30 per ton. See
table, " Iron at $30 per ton ".................................. 3.27

Cost, per pound... c. 4.97

Sell (20 per cent. profit) 6c. per pound.

When cost of moulding is ¾c. per pound.

	Per lb.
Moulding...	c. 0.75
Facings, cores, and chaplets.......................................	.20
Cleaning and chipping...	.20
Labor and handling	.05
Cartage..	.10
Losage on bad castings (10 per cent of above items).................	.13
	c. 1.43

Melted iron. (See table, " Iron at $30 per ton ")................... 3.27

Costs.. c. 4.70

Sell (20 per cent. profit) 5¾c. per pound.

When cost of moulding is 1¼c. per pound.

	Per lb.
Moulding	c. 1.25
Facing, cores, and chaplets..	.20
Cleaning and chipping..........	.20
Labor and handling...	.05
Cartage..	.10
Losage on bad castings (10 per cent. of above items)...............	.18
	c. 1.89

Melted iron. (See table, " Iron at $30 per ton ")................... c. 3 27

Costs.. c. 5.25

Sell (20 per cent. profit) 6¼c. per pound.

FINISHING AND ERECTING SHOP.

Shop expenses (as previously stated), per day.....................		$42 73
Wear and tear of machinery, purchase of new small tools, such as drills, chisels, etc....................................		10 00
Wages—1 foreman........................	$6 00	
1 assistant foreman...........................	4 00	
1 weightman.......................	2 50	
15 finishers, at $3.................................	45 00	
20 finishers, at $2.50...........................	50 00	
25 helpers, at $1.50.............................	37 50	
		145 00
Cartages.....................................		5 00
Contingencies...................................		5 00
Cost, per day............................		$207 73

On 60 men (producers) :

Average wages.............................	$2 41	
Shop expenses.............................	1 05	
Say..................................	$3 50 each workman.	

BLACKSMITH SHOP.

Shop expenses (as previously stated), per day.....................		$6 10
Coal, etc..		5 00
Wages—1 foreman............................	$5 00	
2 smiths, $3 each...............................	6 00	
3 smiths, $2.50 each............................	7 50	
6 helpers, $1.75.................................	10 50	
		29 00
Cartages.......................................		3 00
Cost, per day..............................		$43 10

On 12 men (producers) :

Average wages.........................	$2 41	
Shop expenses.........................	1 18	
Each workman..................................	$3 59	

Cost, per day, of blacksmith and helper—thus :

1 blacksmith		$3 00
1 helper		1 75
Shop expenses, two men each, $1.18............................		2 36
		$7 11
Charge, per day.................................		$8 50

Cost, per day, of blacksmith and two helpers—thus :

```
1 blacksmith ...............................................................  $3 00
2 helpers, $1.75 each .....................................................   3 50
Shop expenses, 3 men each, $1.18.......  ...........................   3 54

    Say ........................................................  ...............  $10 00
    Charge, per day...................................  ........  $12 00
```

TABLE.

Cost of wrought bar iron, with cartage and interest added, and wastage allowed.   Showing rate per pound.

Bar iron at $50 per ton.

```
1 ton iron  (2,240 lbs.) ..............................................$50 00
Interest, four months.......................................... ..   1 17
Cartage.......................................................   2 00
                                                            ─────
                                                            $53 17
```

For wastage (into scrap, etc.), allow 5 per cent.   A gross ton (2,240 lbs.) will yield 2,128 lbs. for finished work.   The bar iron will, therefore, cost, per pound .... c. 2.50

Cost of Bar Iron, at $45 to $85, including interest and cartage.

PER TON.				PER POUND.
Cost of Bar Iron Per Ton.	Four Months' Interest.	Cartage.	Total.	Iron.
$45 00	$1 05	$2 00	$48 05	c. 2.26
47 50	1 11	2 00	50 61	2.38
50 00	1 17	2 00	53 17	2.50
52 50	1 22	2 00	55 72	2.62
55 00	1 28	2 00	58 28	2.74
57 50	1 34	2 00	60 84	2.86
60 00	1 40	2 00	63 40	2.98
62 50	1 46	2 00	65 96	3.10
65 00	1 52	2 00	68 52	3.22
67 50	1 57	2 00	71 07	3.34
70 00	1 63	2 00	73 63	3.46
72 50	1 69	2 00	76 19	3.58
75 00	1 75	2 00	78 75	3.70
77 50	1 81	2 00	81 31	3.87
80 00	1 86	2 00	83 86	3.94
82 50	1 92	2 00	86 42	4.06
85 00	1 98	2 00	88 98	4.18

PATTERN SHOP.

Shop expenses (as previously stated) per day......................		$6 11
Wear and tear of machinery, purchase of small hardware, etc.......		5 00
Wages—1 foreman..	$5 00	
" 7 pattern makers .... $3 00 = 21 00		
" 2 pattern makers .... 2 50 = 5 00		
		31 00
Cartages..		2 00
Cost per day, say		$44 00

On 10 men (producers) :

Average wages..	$3 10
Shop expenses..	1 30
Each workman.................................	$4 40

LIST OF MACHINERY, TOOLS, ETC.,

REQUIRED IN THE VARIOUS SHOPS TO BEGIN WORK WITH.

ENGINE ROOM.

160 horse-power engine and boiler...................	$4,500	
Attachments......................................	1,000	
Fan for cupola....................................	200	
Fan for smith's shop.............................	100	
Shafting throughout buildings.....................	3,000	
		8,800

FOUNDRY.

2 cupolas, maximum capacity 20 tons each	$3,750	
4 cranes...	4,000	
Ladles, shovels, bellows, riddles, sieves, etc..........	2,000	
Sand floors......................................	600	
Weights, etc., etc.................................	2,000	
		12,350

ERECTING AND FINISHING SHOPS.

1 overhead crane..................................	$2,500	
1 column-turning lathe............................	3,500	
1 lathe	750	
1 planer...	600	
1 punch...	650	
1 shears...	500	
Vices and small tools.............................	4,500	
6 drillers..	750	
2 emery wheels...................................	250	
Grindstones......................................	150	
		14,150

2

BLACKSMITH SHOP.

6 forges and tools, anvils, etc........................$3,500		
	3,500	

PATTERN SHOP.

1 wood planer.....................................	$800
2 circular saws....................................	500
1 wood-turning lathe..............................	200
1 wood facing lathe...............................	150
1 jig saw...	175
1 band saw..	175
Benches, etc	500
	2,500

MISCELLANEOUS.

2 hoisting derricks................................$500	
2 hand trucks.....................................	200
4 horses and carts, etc............................	1,800
	2,500
Contingencies.....................	1,700
Total......................................	$45,000

ESTIMATED AMOUNT OF ONE YEAR'S BUSINESS AND COST AND PROFIT.

Interest on capital, 7 per cent. on $150,000........................	$10,500
Taxes, insurance, gas, repairs to buildings, etc	4,500
Office expenses	11,000
Engine and boiler (wages and coal excluded)	1,200
Cupola lining, repairs, etc.......................................	700
Tools, small, additions to, etc...................................	3,600
Lumber in flasks, patterns, etc...................................	4,000
Foundry equipments...	5,000
Moulding sand, sea coal, etc.....................................	2,000
Lump coal, 550 tons, at $6..	3,300
Soft coal for blacksmiths, engine coal, etc........................	1,500
Wood.. ...	500
Paints, oils, etc...	2,000
Stable expenses	2,000
Truckages of heavy work...	1,000
Freights, railroad expenses, etc..................................	2,000

IRON.

Pig, 2,400 tons, average, $32.............................$76,800		
Bar and sheet, 400 tons, average, $90..................	36,000	
Rolled beams, 100 tons, average, $135.....................	13,500	
		126,300

WAGES.

Pattern makers, average number, 10 men—average, $3.10, say..	$31 00
Finishers and helpers, average number, 63 men—average, $2.20, say	138 00
Blacksmiths and helpers, average number, 12 men—average, $2.41, say	29 00
Moulders and helpers, average number, 50 men—average, $2.73, say	136 00
Flask carpenters, average number, 2 men—average, $2.50, say..	5 00
Painters, average number, 3 men—average, $2.25, say......	6 75
Cartmen, average number, 4 men—average, $1.87½, say.....	7 50
Common laborers, average number, 4 men—average, $1.50, say..	6 00
Engineer, 1 man—average, $3, say.......................	3 00
Weightman, 1 man—average, $2.75, say...................	2 75

Total number of men, 150; wages per day, say.....$365 00

300 working days in a year makes the total sum paid for wages.....	$109,500
Contingencies	10,000
Total cost...	$300,000

PRICES REALIZED.

For pig iron made into finished castings, 2,400 tons, at $100........	$240,000
For bar and sheet iron, finished work, 400 tons, at $160...........	64,000
For rolled beams, 100 tons, at $175..............................	17,500
	$321,500

PROFIT.

Profit, $21,500 on an investment of $150,000, in addition to 7 per cent. per annum for use of capital.

The generally profitable character of this business has thus been shown. A new establishment, however, for the first few years, will have to put the larger proportion of its profits in additional machinery, testing apparatus, tools, patterns, etc. In this class of manufacture success attends capacity and industry.

From the preceding tables is obtained the following :

TARIFF OF PRICES FOR LABOR AND MATERIALS.

	Cost.	Charge.
Blacksmith (forge) and 1 helper	$7 11	$8 50
" " 2 helpers........................	10 00	12 00
Finishers, machinists.....................................	3 50	5 00
" extra helpers..................................	2 55	3 50
Pattern makers...	4 40	6 00
Laborers...	1 50	2 00
Drill machine and 1 man.............................	3 50	5 00
Small lathe and 1 man	4 00	6 00
Planer and 1 man...	4 00	6 00
Column-turning machine, etc.............................	6 00	10 00

In making out bills for jobbing work, always charge for the following :

Pattern materials,	Cast iron, by weight,	Cartage,	Blacksmiths,
Screws, etc., files,	Wrought iron, by weight,	Pattern makers,	Machine use,
Bolts and nuts,	Labor,	Finishers,	Boxing.

ILLUSTRATIONS OF COST.

It is important that the cost of every article of common manufacture be made up in a book to be kept for that purpose.   From time to time, as variations arise in prices of labor or materials, these costs must be revised.

As illustrations, the costs of a number of leading articles is given in detail.   From these the manner of definitely arriving at the cost of any article will be readily understood.

ROUND COLUMNS.

Cost of ordinary sharp fluted round columns, with curved heads, and cap and base plates.

	4 in.	5 in.	6 in.	7 in.	8 in	9 in.	10 in.
Diameter	4 in.	5 in.	6 in.	7 in.	8 in	9 in.	10 in.
Length	8 ft.	9 ft.	10 ft.	11 ft.	12 ft.	12 ft.	13 ft.
Average cost of pattern	$0 20	$0 20	$0 25	$0 25	$0 30	$0 35	$0 40
Moulding—3 small columns a day— 1 moulder... $3.00 } 1 helper... $1.50 }	1 50	1 80	2 10	2 40	2 70	3 00	3 50
Facing	25	25	30	35	40	45	50
Core—dry sand	75	85	95	1 10	1 20	1 35	1 50
Chaplets, nails, etc	25	25	30	35	40	45	50
Cleaning	35	25	30	30	35	40	45
Chipping	25	40	45	50	60	70	80
Labor, taking out of foundry and loading	35	25	30	33	40	45	55
Painting	40	50	60	70	80	90	1 00
Cartage	40	50	60	75	85	1 00	1 25
Moulding cap and base plates	12	15	15	15	20	20	20
Turning ends in lathe	60	70	80	90	1 00	1 10	1 20
Losage—1 column in 8	49	57	64	72	82	92	1 06
Weight, including weight of cap and base plates at 3.27 per lb.: Pig iron at $30 per ton [see table, Cost of Melted Iron]	*6 54	*8 83	*11 77	*16 19	*21 97	*24 33	*31 30
Total cost	$12 35	$15 50	$19 51	$25 01	$31 99	$35 60	$44 30
Cost per foot	$1 55	$1 72	$1 95	$2 27	$2 67	$2 67	$3 40
	*200 lbs.	*270 lbs	*360 lbs.	*495 lbs.	*672 lbs.	*744 lbs.	*960 lbs.

*These items pay for breaking up in case of loss in casting.
Add a profit of 20 per cent. to get the selling price.

USUAL SIZE OF PLATES.

Diameter of Column.	Cap Plate.	Base Plate.
4 inch	$10 \times 10 \times \frac{7}{8}$	$8 \times 8 \times \frac{7}{8}$
5 "	$12 \times 12 \times 1$	$10 \times 10 \times 1$
6 "	$12 \times 12 \times 1\frac{1}{8}$	$10 \times 10 \times 1\frac{1}{8}$
7 "	$12 \times 12 \times 1\frac{1}{8}$	$10 \times 10 \times 1\frac{1}{8}$
8 "	$14 \times 14 \times 1\frac{1}{8}$	$12 \times 12 \times 1\frac{1}{4}$
9 "	$14 \times 14 \times 1\frac{1}{8}$	$12 \times 12 \times 1\frac{1}{4}$
10 "	$14 \times 14 \times 1\frac{1}{8}$	$12 \times 12 \times 1\frac{1}{4}$

The cores of columns should be made in one piece in length, the castings made of a uniform thickness of metal, straight and reasonably perfect, and the ends turned off true in a lathe. The somewhat common practice of making columns with a greater thickness at the ends, where it is observable, than at any other part, should not be followed. The risk that is thereby assumed is greater than the value of the metal saved warrants.

It is usual to make round columns a little smaller at the upper end than at the middle; this is advantageous in strength and also in appearance; the difference should not exceed one-tenth of the diameter.

It has been a common opinion that long-continued vibration, though very small in amplitude, causes a change in the structure of iron, rendering it more liable to break; this notion has been abandoned by those best informed.

For columns, good strong iron must always be used, and the castings made with at least ordinary care. It is too much to expect that long columns will be cast quite straight. Some defects can be readily detected by inspection. Those having considerable defects in the casting should, of course, be rejected.

CAST IRON HOLLOW COLUMNS.

Weight per lineal foot, in lbs. avoirdupois. Thickness of metal, from one-quarter to two inches.

Thickness of Metal.......		¼″	⅜″	½″	⅝″	¾″	⅞″	1″	1⅛″	1¼″	1½″	1¾″	2″
		lbs.	lbs.	lbs.	lbs.	lbs.	lbs.	lbs.	lbs.	lbs.	lbs.	lbs.	lbs.
Outside Diameter,	2 inch.	4.29	6.0	7.35	....	....	....	....	....	....	....	....	....
"	2½ "	5.52	7.8	10.0	11.5	....	....	....	....	....	....	....	....
"	3 "	6.75	9.7	12.27	14.6	16.57	....	....	....	....	....	....	....
"	3½ "	7.98	11.5	14.72	17.6	20.25	22.6	....	....	....	....	....	....
"	4 "	9.20	13.3	17.18	20.7	23.92	26.8	29.45	....	....	....	...	....
"	4½ "	10.43	15.2	19.64	23.8	27.61	31.1	34.36	37.3	41.5	....	....	....
"	5 "	11.66	17.1	22.1	26.9	31.29	35.4	39.27	42.8	46.6	....	....	....
"	5½ "	12.88	18.9	24.54	30.0	34.97	39.7	44.18	48.3	51.75	....	.. .	....
"	6 "	14.11	20.7	26.99	33.0	38.65	44.1	49.08	53.8	58.	....	....	....
"	6½ "	15.4	22.6	29.45	36.1	42.34	48.3	53.99	59.4	64.75	....	....	....
"	7 "	16.6	24.4	31.9	39.1	46.02	52.6	58.9	64.9	70.5	80.5	....	....
"	7½ "	17.8	26.2	34.4	42.2	49.7	56.9	63.81	70.4	76.3	87.5	....	....
"	8 "	19.0	28.1	36.8	45.3	53.4	61.2	68.7	76.0	82.5	95.4	....	....
"	8½ "	20.1	29.7	39.3	48.1	57.1	65.3	73.6	81.2	89.	103.0	....	....
"	9 "	21.5	31.7	41.7	51.4	60.7	69.8	78.5	87.0	95.1	110.4	....	....
"	9½ "	22.7	33.6	44.5	51.5	64.4	74.1	83.4	92.5	101.2	117.8	123.0	....
"	10 "	23.9	35.4	46.6	57.5	68.1	78.4	88.4	98.0	107.4	125.2	142.0	157.0
"	10½ "	25.2	37.3	49.1	60.6	71.8	82.7	93.3	103.5	113.5	132.5	150.0	167.0
"	11 "	26.4	39.1	51.5	63.7	75.5	87.0	98.2	109.1	119.6	139.9	159.0	177.0
"	11½ "	27.6	41.0	54.0	66.7	79.1	91.3	103.1	114.6	125.8	147.3	168.0	187.0
"	12 "	28.9	42.8	56.4	69.8	82.8	95.6	108.0	120.1	131.9	154.6	176.0	196.0
"	12½ "	....	44.6	58.9	72.9	86.5	99.9	113.0	125.6	138.1	162.0	185.0	206.0
"	13 "	....	46.5	61.4	75.9	90.2	104.2	117.8	131.2	144.2	169.3	193.0	216.0
"	13½ "	...	. .	63.0	78.5	93.9	108.4	122.9	136.7	150.3	176.7	201.4	226.0
"	14 "	....	...	66.3	82.0	97.6	112.6	127.6	142.0	156.4	184.1	210.1	236.0
"	14½ "	....	...	68.7	85.0	101.2	116.8	132.5	148.0	162.6	191.4	218.2	245.0
"	15 "	....	...	71.2	87.5	104.9	121.2	137.5	153.1	168.7	198.8	226.9	255.0
"	15½ "	....	...	73.6	91.1	108.6	125.5	142.4	159.1	174.9	206.2	235.6	265.0
"	16 "	....	...	76.0	94.1	112.3	129.8	147.3	164.2	181.0	213.5	244.2	275.0
"	16½ "	....	...	78.5	97.3	116.0	134.2	152.3	169.7	187.1	220.9	253.0	285.0
"	17 "	....	...	81.0	100.4	119.7	138.4	157.1	175.2	193.3	228.3	261.7	295.0
"	17½ "	....	...	83.4	103.4	123.3	142.7	162.0	180.8	199.6	235.9	270.0	304.0
"	18 "	....	...	85.9	106.5	127.0	146.9	166.9	186.2	205.5	243.3	278.6	314.0
"	18½ "	....	...	88.3	109.5	130.7	151.3	171.8	191.8	211.7	250.6	287.3	324.0
"	19 "	....	...	90.0	112.2	134.4	155.6	176.7	197.2	217.8	258.0	296.0	334.0
"	19½ "	....	...	93.2	115.6	138.1	159.8	181.6	202.7	223.9	265.4	304.7	344.0
"	20 "	....	...	95.7	118.7	141.7	164.1	186.5	208.4	230.2	272.7	312.8	353.0
"	20½ "	....	...	98.1	121.8	145.4	168.4	191.5	213.9	236.3	280.1	321.6	363.0
"	21 "	....	...	101.7	121.9	149.1	172.8	196.4	219.5	242.6	287.1	330.0	373.0
"	21½ "	....	...	105.3	123.2	152.8	177.0	201.3	225.0	248.7	294.5	338.8	383.0
"	22 "	...	...	108.9	131.3	156.0	181.1	206.2	230.5	254.8	301.8	347.4	393.0
"	22½ "	....	...	111.3	136.0	160.9	186.0	211.1	235.9	260.8	309.2	355.6	402.0
"	23 "	....	...	113.8	139.4	164.9	190.4	216.0	241.5	266.9	316.6	364.3	412.0
"	23½ "	....	...	116.2	142.4	168.6	194.8	220.9	247.0	273.0	323.9	373.0	422.0
Outside Diameter,	24 inch.	....	...	118.7	145.5	172.3	199.0	225.8	252.5	279.1	331.3	381.7	432.0
Thickness of Metal		¼″	⅜″	½″	⅝″	¾″	⅞″	1″	1⅛″	1¼″	1½″	1¾″	2″

Proper allowance must be made for weight of mouldings, ears, or such other projections as may be cast on the columns.

TABLES OF SAFE LOAD ON HOLLOW CYLINDRICAL CAST IRON COLUMNS.

(ONE-FIFTH THE BREAKING WEIGHT.)

The following tables give the safe load in tons of 2,000 pounds, which hollow round iron columns will sustain with safety; the ends turned off true at right angles with their axis; level cap and base plates used; and the columns set up with reasonable care at the building. For columns not turned, one-half of these amounts should be taken for the safe load.

In estimating the load to be borne by a column, allowance must be made for any use the building may be put to, and the greatest weight that may come on any one column. Besides an allowance for the floors, and the weight to be placed thereon, one-fourth of the total should be allowed, in addition, to make assurance doubly sure. Great allowance must also be made for columns that are subject to vibrations caused by machinery, or under the floors of armory drill-rooms, or subject to concussion from bodies falling on a floor above, or liable to lateral blows from goods in transmission being thrown against them.

The castings should be made of a uniform thickness of metal, straight and reasonably perfect, and both ends turned off true in a lathe.

SAFE LOAD, IN TONS, FOR CAST IRON COLUMNS.

Length in Feet	3 INCHES				4 INCHES				5 INCHES				6 INCHES				7 INCHES				8 INCHES				9 INCHES			
	½	¾	1	1¼	½	¾	1	1¼	⅝	¾	1	1¼	¾	1	1¼	1½	¾	1	1¼	1½	¾	1	1¼	1½	¾	1	1¼	1½
7	12.8	15.9	17.2	....	24.9	32.9	38.3	41.7	39.5	53.8	65.0	73.8	77.3	96.5	110.3	122.1	102.4	128.7	150.7	169.4	128.3	162.6	193.1	219.5	154.1	197.7	236.6	271.4
8	10.9	13.0	14.0	....	21.7	28.4	38.0	35.8	35.1	47.6	57.3	64.4	69.7	85.7	98.7	108.6	93.6	117.0	136.9	153.5	118.7	150.1	177.7	201.6	144.7	184.5	226.2	252.0
9	8.9	11.0	11.4	....	19.0	24.8	28.7	31.0	31.3	42.3	56.7	56.8	62.8	77.1	88.5	97.3	85.6	106.7	124.6	139.3	109.8	138.5	163.6	185.2	135.0	171.8	204.7	233.9
10	7.5	8.9	9.6	....	17.4	22.0	24.9	26.3	28.0	37.7	45.1	50.4	56.9	69.6	79.6	87.3	78.4	97.5	113.5	126.6	101.5	127.8	150.7	170.2	126.0	160.0	190.3	217.0
11	6.4	7.6	8.1	....	14.8	18.7	21.1	22.4	25.2	33.6	40.3	44.9	51.6	63.0	71.9	78.7	71.8	89.2	103.6	115.3	94.0	118.0	139.0	156.7	117.5	149.0	177.0	201.4
12	5.4	6.6	7.0	....	12.7	16.2	18.2	19.3	22.7	30.5	36.2	40.8	46.9	57.2	65.2	71.2	66.0	81.7	94.8	105.3	87.0	109.2	128.4	144.8	109.6	138.8	164.5	187.0
13	4.8	5.7	6.1	....	11.1	14.1	15.9	16.8	21.0	27.6	32.2	35.2	42.9	52.1	59.3	61.6	60.7	75.1	87.0	96.6	80.7	101.1	118.5	133.2	102.4	129.4	153.2	173.9
14	4.2	5.0	5.4	....	9.8	12.4	14.0	14.9	18.5	24.3	29.3	31.0	38.3	47.6	54.1	58.9	56.0	69.2	80.0	86.6	75.0	93.8	110.0	123.2	95.7	120.4	142.8	161.9
15	3.7	4.8	4.8	....	8.7	11.1	12.5	13.2	16.5	21.6	25.2	27.6	35.8	43.9	49.0	52.6	51.8	61.9	73.8	81.6	69.8	87.1	101.9	114.2	89.5	112.9	133.3	150.9
16	3.4	4.0	4.3	....	7.8	9.9	11.2	11.8	14.8	19.4	20.6	24.7	33.0	39.4	44.0	47.2	48.1	59.2	68.2	75.4	65.0	81.1	94.7	106.1	83.9	105.7	124.6	140.9
17	3.0	3.6	3.9	....	7.0	8.9	10.1	10.7	13.3	17.5	20.4	22.3	29.8	35.5	39.7	42.5	44.6	51.9	63.2	69.8	60.7	75.7	88.3	98.7	78.7	99.0	116.7	131.8
18	2.8	3.3	3.5	....	6.4	8.1	9.1	9.7	12.1	15.9	18.5	20.2	27.0	32.2	36.0	38.6	42.0	50.9	55.8	63.0	56.8	70.7	82.4	92.1	73.9	92.9	109.4	122.5
19	2.5	3.0	3.2	....	5.8	7.4	8.3	8.8	11.0	14.5	17.4	18.4	24.6	29.4	32.8	35.2	38.3	46.4	52.7	57.4	53.2	66.2	77.1	86.1	69.6	87.4	102.7	115.9
20	2.3	2.7	2.9	....	5.3	6.8	7.6	8.1	10.1	13.3	15.4	15.9	22.6	26.9	30.1	32.3	35.1	42.5	48.3	52.6	51.1	62.7	72.1	79.5	65.5	82.3	96.7	108.9
21	2.1	2.5	2.7	....	4.9	6.2	7.0	7.5	9.3	12.2	14.2	15.5	20.8	24.8	27.7	29.7	32.3	39.1	44.5	48.4	47.0	57.7	66.4	73.2	61.8	75.5	91.0	102.6
22	1.9	2.3	2.5	....	4.6	5.8	6.6	6.9	8.6	11.3	12.9	13.8	19.2	22.9	25.6	27.4	29.8	36.1	41.1	44.4	43.9	53.3	61.3	67.6	58.4	71.6	85.9	96.7
23	1.8	2.1	2.3	....	4.2	5.3	6.0	6.4	8.0	10.5	12.2	13.3	17.8	21.2	23.7	25.4	27.5	33.5	38.1	41.5	40.3	49.4	56.8	62.7	55.0	67.5	81.4	88.3
24	1.7	2.0	2.1	....	3.9	5.0	5.6	5.9	7.4	9.7	11.3	12.4	16.6	19.7	22.1	23.7	25.7	31.2	35.4	38.6	37.5	46.0	52.9	58.3	52.0	64.4	74.8	83.3
25	1.6	1.9	2.0	....	3.7	4.6	5.2	5.5	6.9	9.1	10.6	11.5	15.4	18.4	20.6	22.1	24.0	29.1	33.1	36.0	35.0	42.9	49.3	54.4	48.5	60.1	69.9	77.7

SAFE LOAD, IN TONS, FOR CAST IRON COLUMNS.

LENGTH IN FEET.	OUTSIDE DIAMETER 10 INCHES. Thickness in Inches.				OUTSIDE DIAMETER 11 INCHES. Thickness in Inches.				OUTSIDE DIAMETER 12 INCHES. Thickness in Inches.				OUTSIDE DIAMETER 13 INCHES. Thickness in Inches.				OUTSIDE DIAMETER 14 INCHES. Thickness in Inches.				OUTSIDE DIAMETER 15 INCHES. Thickness in Inches.				OUTSIDE DIAMETER 16 INCHES. Thickness in Inches.			
	¾	1	1¼	1½	1	1¼	1½	2	1	1¼	1½	2	1	1¼	1½	2	1	1¼	1½	2	1	1¼	1½	2	1	1¼	1½	2
7	161.6	231.4	280.9	324.2	269.4	325.9	377.6	469.5	305.2	370.8	431.7	510.9	341.3	414.4	465.7	612.7	377.7	461.7	539.9	654.6	413.7	500.1	584.0	756.7	449.8	551.1	648.0	828.6
8	171.1	219.5	263.8	8-8.9	255.1	308.1	356.8	412.2	290.9	352.3	410.2	512.8	327.0	396.3	461.1	587.9	363.1	442.8	518.0	655.9	399.3	487.9	572.2	727.7	435.3	532.1	606.3	799.8
9	160.9	268.2	247.3	2-1.5	241.2	290.8	336.3	415.6	276.6	335.0	389.1	465.0	312.4	378.3	442.5	555.5	348.5	424.4	496.8	627.0	384.4	469.3	550.1	698.4	420.5	511.4	604.3	770.4
10	151.2	198.4	231.6	296.0	227.8	274.3	316.7	390.3	262.7	317.7	368.6	458.3	298.7	363.6	421.5	537.8	333.0	406.3	474.6	598.5	369.7	451.0	528.2	669.3	405.6	486.6	581.8	741.9
11	142.0	181.4	216.9	243.7	214.9	258.4	298.1	366.3	249.2	301.0	348.8	432.9	284.0	348.4	400.6	501.1	310.4	385.5	453.4	570.7	355.1	433.0	506.3	610.9	390.6	477.4	559.8	711.7
12	133.4	170.1	203.1	292.6	202.1	243.5	280.5	343.9	236.1	285.1	330.0	408.6	270.5	330.3	390.5	475.3	305.4	371.1	42 6	543.6	349.6	415.0	485.0	612.8	376.1	459.3	528.0	683.4
13	125.3	159.6	190.3	217.7	191.2	229.4	264.0	322.8	223.9	270.0	312.2	385.7	257.5	310.8	361.8	450.7	291.8	354.3	412.7	517.7	328.6	397.6	464.5	585.9	361.6	441.2	516.7	655.1
14	117.8	140.8	178.4	203.8	180.5	216.2	248.5	303.3	212.9	255.6	296.3	364.1	245.0	295.5	343.7	427.4	278.8	338.9	398.6	493.0	313.0	380.1	444.4	557.7	347.6	428.3	495.9	626.0
15	110.8	140.7	167.5	191.1	170.3	203.9	234.1	285.1	201.2	242.1	279.1	343.9	233.2	281.1	326.5	405.4	266.2	328.7	375.3	460.4	299.9	364.5	425.2	534.9	333.9	416.9	475.9	601.8
16	104.3	132.4	137.3	179.3	160.9	192.2	220.7	268.3	190.8	229.4	264.5	325.0	222.0	267.3	310.3	384.6	254.3	308.0	367.9	446.9	281.2	348.9	406.7	510.9	320.7	370.6	476.6	576.6
17	93.3	124.6	146.0	168.5	152.1	181.7	208.2	252.7	181.1	217.5	250.0	307.4	211.3	254.4	295.0	365.1	242.9	294.2	341.1	425.7	275.1	334.0	389.1	488.1	306.0	374.9	458.0	552.5
18	92.7	117.4	139.3	158.5	143.9	171.7	196.7	23 3	171.9	206.4	237.5	290.9	201.3	242.1	280.5	346.7	232.0	280.9	325.6	405.5	261.6	319.7	372.2	466.5	215.8	359.9	420.1	629.4
19	87.5	110.8	131.3	149.3	135.2	162.3	185.9	225.0	161.3	195.3	225.3	275.6	191.3	230.6	267.0	329.5	221.7	268.0	310.8	3 6 5	252.5	306.2	356.2	445.3	284.1	345.4	403.0	507.3
20	82.7	104.6	124.0	140.8	129.1	153.9	176.0	212.6	153.2	186.0	213.9	261.3	182.6	219.7	254.2	315.3	212.0	256.1	296.7	368.6	242.9	293.3	341.0	425.3	272.9	331.6	376.5	466.3
21	78.3	99.0	117.2	133.0	122.4	145.9	166.7	201.2	147.5	176.9	203.2	247.9	171.4	209.5	242.6	298.2	162.7	244.7	283.5	351.8	232.0	281.0	326.5	407.8	262.1	318.4	406.2	406.2
22	74.2	98.7	110.9	125.8	116.3	138.4	158.1	190.6	140.6	168.3	193.3	225.5	166.5	199.9	231.9	284.1	194.0	234.0	270.9	335.9	222.3	269.3	312.8	390.3	251.9	305.9	356.4	447.2
23	70.4	88.9	105.1	119.1	110.5	131.5	150.1	180.7	134.0	161.3	184.0	224.0	159.0	190.3	220.4	250.7	185.7	224.0	259.1	321.9	213.4	258.3	299.9	373.7	242.2	293.9	342.3	429.1
24	66.9	84.8	99.7	112.9	105.2	125.1	142.7	171.6	127.8	152.9	175.3	213.2	152.0	182.3	210.4	258.3	177.9	214.4	248.0	306.5	204.9	247.8	287.5	358.1	232.9	282.5	328.8	411.9
25	64.9	81.0	94.2	106.3	100.2	119.1	135.7	163.1	122.0	145.6	167.1	203.1	145.1	171.3	201.0	246.6	170.4	205.4	237.5	294.1	196.7	237.8	275.9	348.2	224.0	271.6	316.1	395.6

ROUND COLUMNS.

Cost of deep fluted round columns with full leaf Corinthian capitals and Attic bases.

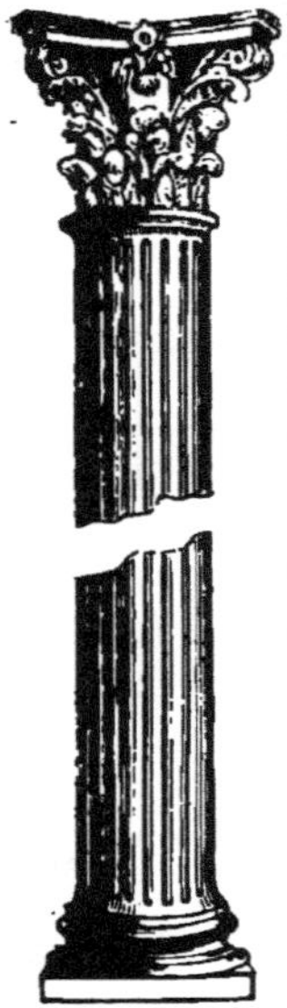

Diameter	7 inch.		8 inch.		10 inch.		12 inches.		14 inches.		16 inches.	
Length	12 feet.		12 feet.		12 feet.		12 feet.		12 feet.		12 feet.	
Average cost of pattern	$0·40		$0 50		$0 70		$0 85		$1 00		$1 25	
Moulding { ¾ day moulder $3.00, ¾ day helper 1.50 }	3 38		*4 50		*5 38		*6 75		*7 88		*9 00	
Facing	38		42		50		60		70		85	
Core	1 10		1 25		1 50		1 75		2 00		2 25	
Chaplets, nails, etc.	38		42		50		55		60		65	
Cleaning	38 ‡		45 ‡		50 ‡		70 ‡		85 ‡		1 00 ‡	
Chipping	55 ‡		60 ‡		85 ‡		1 20 ‡		1 50 ‡		1 75 ‡	
Labor, taking out of foundry and loading	35		40		50		75		1 20		1 50	
Painting	75		85		1 00		1 10		1 20		1 30	
Cartage	75		75		1 00		1 00		1 25		1 25	
Moulding cap and base plates	20		20		24		24		30		30	
Turning ends in lathe	90		90		1 25		1 22		1 60		1 60	
Losage, 1 column in 8	69		1 09		1 33		1 67		2 00		2 32	
Weight (including weight of cap and base plates, at c. 3.27 per lb.	†19 62		†24 33		†33 35		†39 24		†46 09		†55 93	
[Pig iron taken at $30 per ton. See table of Cost of Melted Iron.]		$30 23		$26 66		$48 60		$57 65		$68 17		$80 95
			*1 day.		*1¼ days.		*1½ days.		*1¾ days.		*2 days.	
	†600 lbs.		†744 lbs.		†1020 lbs.		†1200 lbs.		†1440 lbs.		†1680 lbs.	
COST OF CAPITAL.												
Outside diameter at neck	5⅜ in.		6¼ in.		8⅛ in.		10 in.		12 in.		14 in.	
Weight, at c. 3.27 per lb.	$0 98		*$1 30		*$1 96		*$2 43		*3 43		*$4 58	
Moulding	1 50		1 75		2 00		2 20		2 60		3 00	
Other expenses in foundry	30		50		65		80		95		1 20	
Finishing, fitting and putting up	2 50		2 75		3 00		3 50		4 00		4 50	
Screws, rivets, files, etc.	25		30		35		40		45		50	
Painting	25		40		45		50		60		75	
Cartage	10		15		20		25		35		50	
		5 98		7 15		8 61		10 08		12 38		15 03
	*30 lbs.		*40 lbs.		*60 lbs.		*75 lbs.		*105 lbs.		*140 lbs.	
COST OF SHELL BASE.												
Weight, at c. 3.27 per lb.	*$0 91		*$0 98		*$1 14		*$1 30		*$1 64		*$2 13	
Moulding	50		65		75		90		1 05		1 20	
Other expenses in foundry	20		30		25		40		50		60	
Finishing, fitting and putting up	40		45		50		60		70		80	
Screws, files, etc.	12		15		18		22		26		70	
Painting	10		12		15		20		30		40	
Cartage	10		12		15		18		20		22	
		2 33		2 77		3 22		3 80		4 65		5 63
	*28 lbs.		*40 lbs.		*35 lbs.		*40 lbs.		*50 lbs.		*65 lbs.	
Total cost		$38 54		$46 58		$60 43		$71 53		$85 20		$101 61

‡ These items pay for breaking up in case of loss in casting.
Add a profit of 20 per cent. to get the selling price.
NOTE.—It will be observed that when the length of the column is increased or diminished, the cost varies only as to the quantity of iron in the column. All the other items remain the same.

BOX COLUMNS.

Cost of panelled box columns, including the cost of workmanship and materials of capitals, and fastening the same on at building.

Size.	26×12 and 8×12.	8×14 and 8×16.	10×12 and 12×12	12×18 and 12×20	16×18 and 14×20
Average cost of patterns	$1 00	$1 00	$1 25	$1 25	$1 50
Moulding	4 50	5 00	5 60	6 00	6 75
Facing	50	60	70	80	90
Core	1 00	1 00	1 20	1 30	1 40
Chaplets, nails, etc.	25	25	30	30	35
Cleaning	40 *	50 *	60 *	75 *	1 00 *
Chipping	1 00	1 00	1 15	1 25	1 50
Labor, getting in flasks, taking out casting, etc.	1 00	1 15	1 30	1 50	1 75
Painting	75	80	85	90	1 00
Cartage	1 00	1 00	1 25	1 25	1 50
Moulding cap and base plates	20	22	24	26	28
Turning ends in lathe	1 00	1 00	1 20	1 20	1 40
Losage, 1 column in 8	1 23	1 35	1 53	1 63	1 93
	$13 83	$14 97	$17 07	$18 44	$21 96

Cost of Corinthian capital running back the full depth of the column on both sides.

	26×12 and 8×12.	8×14 and 8×16.	10×12 and 12×12	12×18 and 12×20	16×18 and 14×20
Weight, at c. 3.27 per lb.	2 19†	2 84†	3 27†	3 92†	4 64†
Moulding	1 80	2 10	2 40	2 70	3 00
Other expenses in foundry	60	80	1 00	1 90	1 40
Finishing, fitting and putting up	4 00	4 00	4 25	4 25	4 50
Screws, files, etc.	25	25	25	30	35
Painting	25	25	30	30	25
Cartage	15	15	20	20	25
	$9 24	$10 30	$11 67	$12 87	$14 49
The average for windows is two-thirds of this	$6 16	$6 93	$7 78	$8 58	$9 66
Cost, without iron (except the weight of capital)	$19 99	$21 90	$24 85	$27 02	$30 92
	167 lbs.	167 lbs.	100 lbs.	120 lbs.	142 lbs.

* These items pay for breaking up in case of loss in casting.

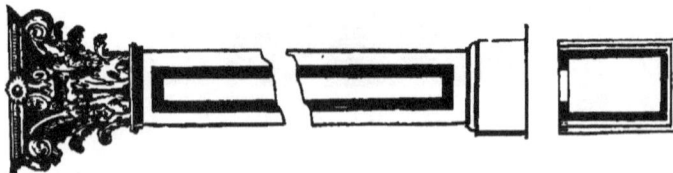

TABLE.

[Arranged from the foregoing details.]

COST OF WORKMANSHIP ON BOX COLUMNS, INCLUDING LEAF CAPITALS, SAY :

6 × 12 ....$14 00	10 × 12.....$17 00	14 × 12.....$19 00	18 × 12....$20 00
6 × 14 .... 14 50	10 × 14..... 17 50	14 × 14..... 19 50	18 × 14 ... 20 50
6 × 16 .... 15 00	10 × 16. ... 18 00	14 × 16..... 20 00	18 × 16.... 21 00
6 × 18 .... 15 50	10 × 18..... 18 50	14 × 18..... 20 50	18 × 18.... 22 00
6 × 20 .... 16 00	10 × 20..... 19 00	14 × 20..... 21 00	18 × 20.... 23 00
8 × 12 .... 14 00	12 × 12..... 17 00	16 × 12..... 19 00	20 × 12.... 20 00
8 × 14 .... 14 50	12 × 14..... 17 50	16 × 14..... 19 50	20 × 14.... 21 00
8 × 16 .... 15 00	12 × 16..... 18 00	16 × 16..... 20 00	20 × 16.... 22 00
8 × 18 .... 15 50	12 × 18..... 18 50	16 × 18..... 21 00	20 × 18.... 23 00
8 × 20 .... 16 00	12 × 20..... 19 00	16 × 20..... 22 00	20 × 20.... 24 00

TABLE.

WEIGHTS OF ORDINARY BOX COLUMNS, PANELLED, MADE AS LIGHT AS CAN BE SAFELY RUN, AND WITH OPEN BACKS.

[Plates included.]

Size.	Weight in lbs.	Size.	Weight in lbs.	Size.	Weight in lbs.	Size.	Weight in lbs.
6 × 10.......	69	10 × 10........	87	14 × 10.......	105	18 × 10.......	126
6 × 12.......	78	10 × 12........	96	14 × 12.......	113	18 × 12.......	133
6 × 14.......	89	10 × 14........	105	14 × 14.......	123	18 × 14.......	143
6 × 16.......	99	10 × 16........	113	14 × 16	134	18 × 16.......	154
6 × 18......	109	10 × 18........	126	14 × 18.......	145	18 × 18.......	166
6 × 20......	120	10 × 20........	138	14 × 20.. ...	160	18 × 20.......	180
8 × 10.......	78	12 × 10........	96	16 × 10.......	113	20 × 10.......	138
8 × 12.......	87	12 × 12........	104	16 × 12.......	123	20 × 12.......	146
8 × 14.......	96	12 × 14........	113	16 × 14.......	133	20 × 14.......	153
8 × 16......	107	12 × 16........	123	16 × 16.......	144	20 × 16.......	165
8 × 18......	118	12 × 18........	136	16 × 18.......	156	20 × 18.......	178
8 × 20......	128	12 × 20........	149	16 × 20.......	170	20 × 20.......	190

What is the cost of a box column 14 inches face, 16 inches deep, and 12 feet long?

Weight, 134 lbs. to a foot [see above table] = 1608 lbs. @ c. 3.27..... $52 50
Workmanship, including the capital............................... 20 08

   Cost.. $72 58
Add 20 per cent. profit.. ... 14 52

Sell.. $87 10
   Is $7.27 per lineal foot, or c. 5.42 per pound.

NOTE.—If columns are deep panelled or heavy mouldings in panels, the weight will be considerably more.

If shutter grooves are required, add for additional weight and labor. Setting of columns always charged in addition.

CAST IRON BEAMS.

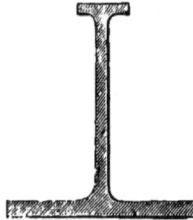

The best form of section for cast iron beams or girders is that known as T beams. Experiment has established the rule, that the area of the bottom flange should be a little more than six times that of the top flange, and the flanges connected together by a vertical web curved in the shape of an ellipse, and sufficiently rigid to give lateral stiffness.

A cast iron beam will be bent to one-third of its breaking weight if the load is laid on gradually; and one-sixth of it, if laid on at once, will produce the same effect if the weight of the beam is small compared with the weight laid on. Hence the breaking weight of the beam should not be less than three times the greatest load which it has to carry, and for those exposed to vibrations the strength should not be less than six times the load imposed, as sudden shocks tend far more to destroy the cohesion than a permanent load.

RULE.

The rule to determine the strength of such beams is as follows:

Multiply the sectional area of the bottom flange in inches by the depth of the beam in inches, and divide the product by the distance between the supports, also in inches; and 514 times the quotient equals the absolute strength of the beam in cwts.

EXAMPLE.

What is the load that will break a T beam of the following dimensions: ten feet in length between supports, the load applied in the middle?

Top flange......... $7'' \times 1''$.
Centre web.........$21'' \times \frac{8}{10}''$ — $15''$ at ends.
Bottom flange.......$21'' \times 2''$.

As per rule:
$21'' \times 2'' = 42'' \times 21'' = 882'' \div 120'' = 7.35'' \times 514 = 3,778$ cwt. or $188\frac{18}{20}$ tons.

A cubic foot of brick work weighs 112 pounds—a foot of wall sixteen inches thick will weigh 150 pounds.

In the following tables the thicknesses for the castings are set forth, and other necessary details. The weight of brick-work is calculated as a solid wall equally distributed, exclusive of floors or any other weight. If window openings occur, deduct only half weight; that is, take out of the weight of wall only half the actual space which the windows will occupy. Should the weight of wall, by piers or otherwise, be placed at or near the centre of the girder, double the weight calculated to be borne; in other words, use a girder of greater sustaining capacity.

TABLE.

I BEAMS TO SUSTAIN TWO STORIES, OR TWENTY-SIX FEET HIGH, OF TWELVE-INCH BRICK WALL.

Whole Length of Girder.	Bearing on Wall at each End.	Distance between Supports.	Top Flange.	Centre Web.	Bottom Flange.	Weight of Casting.	Weight of Brick Wall.
7 feet 0 inches.	6 inches.	6 feet.	1¼ × 1 inch.	Centre 10″ × ⅝″ — 6″ at ends.	10 × ¾ inches.	301 lbs.	Say 9 net tons.
8 " 2 "	7 "	7 "	1½ × 1 "	" 10 × ⅝ — 6 "	10 × ⅞ "	392 "	" 10½ "
9 " 4 "	8 "	8 "	1¾ × 1 "	" 11 × ⅝ — 6½ "	10 × 1 "	495 "	" 12 "
10 " 6 "	9 "	9 "	2 × 1 "	" 12 × ⅝ — 7 "	12 × ⅞ "	588 "	" 13½ "
11 " 6 "	9 "	10 "	2 × 1 "	" 13 × ⅝ — 7½ "	12 × 1 "	725 "	" 15 "
12 " 6 "	9 "	11 "	2¼ × 1 "	" 14 × ⅝ — 8 "	12 × 1⅛ "	875 "	" 16½ "
13 " 8 "	10 "	12 "	2¼ × 1 "	" 15 × ¾ — 9 "	12 × 1⅓ "	1039 "	" 18 "
14 " 8 "	10 "	13 "	2½ × 1 "	" 16 × ¾ — 10 "	12 × 1¼ "	1232 "	" 19½ "
16 " 0 "	12 "	14 "	2¾ × 1 "	" 17 × ⅞ — 10 "	12 × 1⅜ "	1520 "	" 21 "
17 " 0 "	12 "	15 "	3 × 1 "	" 18 × 1 — 11 "	12 × 1½ "	1836 "	" 22½ "
18 " 4 "	14 "	16 "	3 × 1 "	" 19 × 1$\frac{1}{16}$ — 12 "	12 × 1½ "	2127 "	" 24 "
19 " 4 "	14 "	17 "	3¼ × 1 "	" 20 × 1⅓ — 13½ "	12 × 1¾ "	2417 "	" 25½ "
20 " 6 "	15 "	18 "	3¾ × 1 "	" 21 × 1⅛ — 14 "	12 × 1¾ "	2786 "	" 27 "
21 " 8 "	16 "	19 "	4 × 1 "	" 21 × 1¼ — 14 "	12 × 1⅞ "	3186 "	" 28½ "
22 " 8 "	16 "	20 "	4 × 1 "	" 22½ × 1¼ — 15 "	12 × 2 "	3432 "	" 30 "

TABLE.

BEAMS TO SUSTAIN THREE STORIES, OR THIRTY-EIGHT FEET HIGH, OF TWELVE-INCH BRICK WALL.

Whole Length of Girder.	Bearing on Wall at each End.	Distance between Supports.	Top Flange.	Centre Web.	Bottom Flange.	Weight of Casting.	Weight of Brick Wall.
8 feet 6 inches.	9 inches.	7 feet.	2" × 1 inch.	Centre 10" × ¾" — 6" at ends.	12" × 1 inches.	527 lbs.	Say 14 net tons.
9 " 6 "	9 "	8 "	2 × 1¼ "	" 12 × ¾ — 7½ "	12 × 1⅛ "	684 "	" 17 "
10 " 6 "	9 "	9 "	3 × ⅞ "	" 14 × ¾ — 8½ "	12 × 1$\frac{3}{16}$ "	830 "	" 19 "
11 " 8 "	10 "	10 "	3 × ⅞ "	" 14 × ⅞ — 8½ "	12 × 1¼ "	980 "	" 21 "
12 " 8 "	10 "	11 "	3 × 1 "	" 15 × ⅞ — 9 "	12 × 1$\frac{5}{16}$ "	1140 "	" 23 "
13 " 8 "	10 "	12 "	2½ × 1¼ "	" 15 × 1 — 9 "	12 × 1½ "	1381 "	" 25 "
15 " 0 "	12 "	13 "	2½ × 1½ "	" 16 × 1$\frac{1}{16}$ — 10 "	12 × 1⅜ "	1680 "	" 27 "
16 " 0 "	12 "	14 "	3½ × 1 "	" 17 × 1⅛ — 11 "	12 × 1¾ "	1952 "	" 29 "
17 " 4 "	14 "	15 "	4 × 1⅛ "	" 18½ × 1$\frac{1}{16}$ — 12 "	12 × 1⅞ "	2392 "	" 31 "
18 " 6 "	15 "	16 "	4 × 1⅛ "	" 20 × 1¼ — 13 "	12 × 2 "	2757 "	" 34 "
19 " 6 "	15 "	17 "	4 × 1¼ "	" 20 × 1¼ — 13 "	12 × 2⅛ "	3042 "	" 36 "
20 " 6 "	15 "	18 "	4 × 1¼ "	" 21 × 1⅜ — 14 "	12 × 2¼ "	3485 "	" 38 "
21 " 8 "	16 "	19 "	4 × 1⅜ "	" 22½ × 1⅜ — 14½ "	12 × 2⅜ "	3830 "	" 40 "
22 " 8 "	16 "	20 "	4 × 1⅜ "	" 23 × 1½ — 15 "	12 × 2⅜ "	4262 "	" 42 "

3

TABLE.

⊥ BEAMS TO SUSTAIN FOUR STORIES, OR FIFTY FEET HIGH, OF TWELVE-INCH BRICK WALL.

Whole Length of Girder.	Bearing on Wall at each End.	Distance between Supports.	Top Flange.	Centre Web.	Bottom Flange.	Weight of Casting.	Weight of Brick Wall.
8 feet 6 inches.	9 inches.	7 feet.	2" × 1¼ inch.	Centre 12" × ¾ — 7" at ends.	12" × 1⅛ inches.	604 lbs.	Say 20 net tons.
9 " 6 "	9 "	8 "	2½ × 1 "	" 14 × ¾ — 8½ "	12 × 1 3/16 "	732 "	" 22 "
10 " 8 "	10 "	9 "	3 × 1 "	" 15 × ⅞ — 9 "	12 × 1¼ "	933 "	" 25 "
11 " 8 "	10 "	10 "	2¼ × 1¼ "	" 15 × 1 — 9 "	12 × 1⅜ "	1132 "	" 28 "
12 " 8 "	10 "	11 "	2½ × 1¼ "	" 15 × 1⅛ — 9 "	12 × 1½ "	1272 "	" 30 "
13 " 8 "	10 "	12 "	2½ × 1½ "	" 16 × 1 3/16 — 10 "	12 × 1⅝ "	1503 "	" 33 "
15 " 0 "	12 "	13 "	3 × 1¼ "	" 17 × 1¼ — 11 "	12 × 1¾ "	1920 "	" 36 "
16 " 0 "	12 "	14 "	3 × 1½ "	" 18 × 1¼ — 12 "	12 × 2 "	2256 "	" 39 "
17 " 4 "	14 "	15 "	3 × 1½ "	" 18½ × 1 5/16 — 12½ "	12 × 2⅛ "	2565 "	" 42 "
18 " 4 "	14 "	16 "	4 × 1¼ "	" 19½ × 1 5/16 — 12½ "	12 × 2¼ "	2933 "	" 44 "
19 " 6 "	15 "	17 "	4 × 1¼ "	" 21 × 1⅜ — 13½ "	12 × 2⅜ "	3374 "	" 47 "
20 " 6 "	15 "	18 "	4 × 1¼ "	" 23 × 1⅜ — 14 "	12 × 2½ "	3772 "	" 50 "
21 " 6 "	15 "	19 "	4 × 1⅜ "	" 23 × 1½ — 15 "	12 × 2⅝ "	4236 "	" 53 "
22 " 8 "	16 "	20 "	4 × 1½ "	" 25 × 1½ — 15 "	12 × 2¾ "	4715 "	" 56 "

TABLE.

I BEAMS TO SUSTAIN THREE STORIES, OR THIRTY-EIGHT FEET HIGH, OF SIXTEEN-INCH BRICK WALL.

Whole Length of Girder.	Bearing on Wall at each End.	Distance between Supports.	Top Flange.	Centre Web.	Bottom Flange.	Weight of Casting.	Weight of Brick Wall.
9 feet 6 inches.	9 inches.	8 feet.	2″ × 1½ inch.	Centre 12″ × ¾″ — 7″ at ends.	16″ × 1⅛ inches.	817 lbs.	Say 22 net tons.
10 " 8 "	10 "	9 "	2 × 1½ "	" 13 × $1\frac{3}{16}$ — 8 "	16 × 1¼ "	1045 "	" 25 "
11 " 8 "	10 "	10 "	3 × 1¼ "	" 14 × ⅞ — 8 "	16 × 1⅜ "	1272 "	" 28 "
13 " 0 "	12 "	11 "	3 × 1¼ "	" 15 × ⅞ — 9 "	16 × 1⅜ "	1469 "	" 31 "
14 " 0 "	12 "	12 "	3 × 1¼ "	" 15 × 1 — 9 "	16 × 1½ "	1708 "	" 34 "
15 " 0 "	12 "	13 "	3 × 1½ "	" 16 × $1\frac{1}{16}$ —10 "	16 × 1⅝ "	2085 "	" 37 "
16 " 0 "	12 "	14 "	3 × 1½ "	" 17 × 1⅛ —11 "	16 × 1¾ "	2416 "	" 39 "
17 " 4 "	14 "	15 "	4 × 1¼ "	" 18 × 1¼ —13 "	16 × 1⅞ "	2895 "	" 42 "
18 " 6 "	15 "	16 "	4 × 1¾ "	" 19 × 1⅜ —13½ "	16 × 2 "	3367 "	" 45 "
19 " 6 "	15 "	17 "	4 × 1½ "	" 20 × 1⅜ —14 "	16 × 2⅛ "	3803 "	" 48 "
20 " 8 "	16 "	18 "	4 × 1½ "	" 20 × 1⅜ —14 "	16 × 2¼ "	4175 "	" 51 "
21 " 8 "	16 "	19 "	4½ × 1½ "	" 22 × 1⅜ —15 "	16 × 2⅜ "	4745 "	" 54 "
22 " 8 "	16 "	20 "	5 × 1½ "	" 22 × 1½ —15 "	16 × 2½ "	5327 "	" 57 "

TABLE.

⊥ BEAMS TO SUSTAIN FOUR STORIES, OR FIFTY FEET HIGH, OF SIXTEEN-INCH BRICK WALL.

Whole Length of Girder.	Bearing on Wall at each End.	Distance between Supports.	Top Flange.	Centre Web.	Bottom Flange.	Weight of Casting.	Weight of Brick Wall.
8 feet 6 inches.	9 inches.	7 feet.	2″ × $1\frac{1}{2}$ inch.	Centre 12″ × $\frac{13}{16}$ — 7″ at ends.	16″ × $1\frac{1}{8}$ inches.	765 lbs.	Say 25 net tons.
9 " 8 "	10 "	8 "	$2\frac{1}{4}$ × $1\frac{1}{4}$ "	" 14 × $\frac{7}{8}$ — 8 "	16 × $1\frac{1}{4}$ "	1007 "	" 29 "
11 " 0 "	12 "	9 "	3 × $1\frac{1}{4}$ "	" 15 × $\frac{7}{8}$ — 9 "	16 × $1\frac{3}{8}$ "	1243 "	" 32 "
12 " 0 "	12 "	10 "	3 × $1\frac{1}{2}$ "	" 15 × $\frac{15}{16}$ — 9 "	16 × $1\frac{1}{2}$ "	1476 "	" 36 "
13 " 0 "	12 "	11 "	3 × $1\frac{1}{2}$ "	" 16 × 1 —10 "	16 × $1\frac{5}{8}$ "	1742 "	" 40 "
14 " 0 "	12 "	12 "	4 × $1\frac{1}{4}$ "	" 17 × $1\frac{1}{16}$ —11 "	15 × $1\frac{3}{4}$ "	2072 "	" 44 "
15 " 4 "	14 "	13 "	4 × $1\frac{1}{4}$ "	" 18 × $1\frac{1}{8}$ —12 "	16 × $1\frac{7}{8}$ "	2469 "	" 47 "
16 " 4 "	14 "	14 "	4 × $1\frac{1}{2}$ "	" 19 × $1\frac{1}{8}$ —13 "	16 × 2 "	2858 "	" 51 "
17 " 6 "	15 "	15 "	4 × $1\frac{1}{2}$ "	" 20 × $1\frac{1}{4}$ —$13\frac{1}{2}$ "	16 × $2\frac{1}{8}$ "	3255 "	" 55 "
18 " 6 "	15 "	16 "	$4\frac{1}{2}$ × $1\frac{1}{2}$ "	" 21 × $1\frac{3}{8}$ —14 "	16 × $2\frac{1}{4}$ "	3922 "	" 59 "
19 " 6 "	15 "	17 "	5 × $1\frac{1}{2}$ "	" 23 × $1\frac{1}{2}$ —15 "	16 × $2\frac{1}{2}$ "	4680 "	" 62 "
20 " 8 "	16 "	18 "	5 × $1\frac{1}{2}$ "	" 25 × $1\frac{1}{2}$ —16 "	16 × $2\frac{5}{8}$ "	5229 "	" 66 "

EXAMPLE OF COST OF CAST IRON BEAM.

Suitable to sustain three stories of twelve-inch brick wall. Length between supports, eighteen feet.

Top flange, 4″ × 1¼″.
Centre web, 21 × 1⅜—14 inches at ends.
Bottom flange, 12 ″ × 2¼″.

Average cost of pattern, flasks, etc.		$2 00
Moulding—2 moulders 1 day, $3.00.	$6 00	
1 helper 1 day.	1 50——	7 50
Facing		1 00
Chaplets, etc.		40
Cleaning.		50
Chipping		1 50
Labor, bringing in flasks, getting out castings, etc.		1 00
Sundries		1 00
Painting.		1 25
Cartage		3 00
Losage—1 beam in 15.		2 24
Weight, 3,485 lbs. @ c. 3.27.		113 96
Cost		$135 35

Add 20 per cent. profit for selling price.

CAST IRON ARCH GIRDERS, WITH WROUGHT IRON TENSION RODS.

Arch girders are principally used for the support of the front or rear walls of brick buildings. They are a cheap and effective method of securing wide openings. The casting is made in one piece with box ends, the latter having grooves and seats to receive the wrought iron tie rod. The tie rod is made from one-eighth to three-eighths of an inch shorter than the casting; and has square ends forming shoulders so as to fit into the casting. The rod has usually one weld on its length, and great care should be taken that this weld be perfect.

The rod is expanded by heat, and then placed in position in the casting, and allowed to contract in cooling, thus tieing the two ends of the casting together to form abutments to receive the horizontal thrust of the arch. If the tie rod is too long it will not receive the full proportion of the strain until the cast iron has so far deflected that its lower edge is subjected to a severe tensile strain, which cast iron is feeble to resist. If the tie rod is made too short, the casting is cambered up and a severe initial strain put upon both the cast and wrought iron, which enfeebles both in carrying a load. The proper proportion of cast iron arch to wrought iron tie; the proper welding and shrinkage of the bar are all important elements. The girders should have a rise of about two feet six inches on a length of twenty-five feet. One square inch of cross-section of rod should be allowed for every ten net tons of load imposed upon the span of the arch.

In the following tables the thickness for the castings are set forth, the proper diameter of tie rods to be used, and other necessary details. The weight of brick work is calculated as a solid wall equally distributed, exclusive of floors or any other weight. If window openings occur, deduct only half weight; that is, take out of the weight of wall only half the actual space which the windows will occupy. Should the weight of wall, by piers or otherwise, be placed at or near the centre of the girder, double the weight calculated to be borne; in other words, use a girder of greater sustaining capacity.

A cubic foot of brick work weighs 112 pounds; a foot of wall sixteen inches thick will weigh 150 pounds.

TO SUSTAIN 3 STORIES, OR 40 FEET HIGH, OF 12-INCH BRICK WALL.

SECTION OF GIRDER.

TOP FLANGE.... 12" × 1"

CENTRE WEB.... 12" × ¼"

BULB.......... 3" × 2"

Whole Length of Girder.	Bearing on Wall at Each End.	Span.	Size of Rod.	Weight of Rod.	Weight of Casting.	Weight of Brick Work.
16 feet.	Say 1 foot 6 inches.	13 feet.	1¼ inches Round.	138 lbs.	1,500 lbs.	Say 20 Net Tons.
17 "	" 1 " 6 "	14 "	1¼ "	145 "	1,580 "	" 31 "
18 "	" 1 " 6 "	15 "	2 "	202 "	1,660 "	" 33 "
19 "	" 1 " 6 "	16 "	2 "	212 "	1,740 "	" 35 "
20 "	" 1 " 6 "	17 " 4 inches.	2 "	223 "	1,820 "	" 37 "
21 "	" 1 " 10 "	17 " 4 "	2¼ "	297 "	1,900 "	" 39 "
22 "	" 1 " 10 "	18 " 4 "	2¼ "	311 "	1,980 "	" 41 "
23 "	" 1 " 10 "	19 " 4 "	2¼ "	324 "	2,080 "	" 43 "
24 "	" 1 " 10 "	20 " 4 "	2¼ "	338 "	2,160 "	" 45 "
25 "	" 1 " 10 "	21 " 4 "	2¼ "	434 "	2,240 "	" 47 "
26 "	" 2 feet "	22 "	2¼ "	451 "	2,320 "	" 50 "
27 "	" 2 "	23 "	2¼ "	560 "	2,480 "	" 52 "
28 "	" 2 "	24 "	3 "	696 "	2,560 "	" 54 "
29 "	" 2 "	25 "	3 "	720 "	2,640 "	" 56 "
30 "	" 2 "	26 "	3 "	743 "	2,720 "	" 58 "

☞ NOTE.—When Iron Columns support the Girders, the bearings will be less.

TO SUSTAIN 4 STORIES, OR 50 FEET HIGH, OF 12-INCH BRICK WALL.

SECTION OF GIRDER.

TOP FLANGE..... 12" × 1¼"

CENTRE WEB.... 12" × ⅞"

BULB........... 3" × 2"

Whole Length of Girder.	Bearing on Wall at Each End.	Span.	Size of Rod.	Weight of Rod.	Weight of Casting.	Weight of Brick Work.
16 feet.	Say 1 foot 6 inches.	13 feet.	2 inches Round.	180 lbs.	1,800 lbs.	Say 36 Net Tons.
17 "	" 1 " 6 "	14 "	2 "	190 "	1,900 "	" 39 "
18 "	" 1 " 6 "	15 "	2 "	201 "	2,000 "	" 42 "
19 "	" 1 " 6 "	16 "	2¼ "	270 "	2,100 "	" 45 "
20 "	" 1 " 6 "	17 "	2¼ "	283 "	2,200 "	" 48 "
21 "	" 1 " 10 "	17 " 4 inches.	2¼ "	297 "	2,300 "	" 51 "
22 "	" 1 " 10 "	18 " 4 "	2½ "	384 "	2,400 "	" 54 "
23 "	" 1 " 10 "	19 " 4 "	2½ "	410 "	2,500 "	" 56 "
24 "	" 1 " 10 "	20 " 4 "	2½ "	417 "	2,600 "	" 59 "
25 "	" 1 " 10 "	21 " 4 "	2¾ "	520 "	2,700 "	" 61 "
26 "	" 2 feet.	22 "	2¾ "	540 "	2,800 "	" 63 "
27 "	" 2 "	23 "	3 "	672 "	2,900 "	" 65 "
28 "	" 2 "	24 "	3 "	693 "	3,000 "	" 68 "
29 "	" 2 "	25 "	3¼ "	830 "	3,100 "	" 71 "
30 "	" 2 "	26 "	3¼ "	857 "	3,200 "	" 74 "

☞ NOTE.—When Iron Columns support the Girders, the bearings will be less.

TO SUSTAIN 3 STORIES, OR 40 FEET HIGH, OF 16-INCH BRICK WALL.

SECTION OF GIRDER.

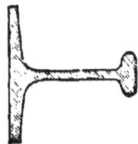

TOP FLANGE.... 12" × 1¼"

CENTRE WEB.... 12" × ¾"

BULB......... 3½" × 2"

Whole Length of Girder.	Bearing on Wall at Each End.	Span.	Size of Rod.	Weight of Rod.	Weight of Casting.	Weight of Brick Work.
16 feet.	Say 1 foot 6 inches.	13 feet.	2 inches Round.	180 lbs.	2,000 lbs.	Say 39 Net Tons.
17 "	" 1 " 6 "	14 "	2 "	190 "	2,125 "	" 42 "
18 "	" 1 " 6 "	15 "	2 "	201 "	2,250 "	" 45 "
19 "	" 1 " 6 "	16 "	2¼ "	270 "	2,375 "	" 48 "
20 "	" 1 " 6 "	17 "	2¼ "	283 "	2,500 "	" 51 "
21 "	" 1 " 10 "	17 " 4 inches.	2¼ "	297 "	2,625 "	" 52 "
22 "	" 1 " 10 "	18 " 4 "	2¼ "	384 "	2,750 "	" 55 "
23 "	" 1 " 10 "	19 " 4 "	2¼ "	410 "	2,875 "	" 58 "
24 "	" 1 " 10 "	20 " 4 "	2¼ "	437 "	3,000 "	" 61 "
25 "	" 1 " 10 "	21 " 4 "	2¼ "	520 "	3,125 "	" 64 "
26 "	" 2 feet.	22 "	3 "	540 "	3,250 "	" 66 "
27 "	" 2 " "	23 "	3 "	672 "	3,375 "	" 69 "
28 "	" 2 " "	24 "	3 "	693 "	3,500 "	" 72 "
29 "	" 2 " "	25 "	3¼ "	830 "	3,625 "	" 75 "
30 "	" 2 " "	26 "	3¼ "	857 "	3,650 "	" 78 "

☞ NOTE.—When Iron Columns support the Girders, the bearings will be less.

TO SUSTAIN 4 STORIES, OR 50 FEET HIGH, OF 16-INCH BRICK WALL.

SECTION OF GIRDER.

TOP FLANGE..... 16" × 1½"

CENTRE WEB.... 12" × 1"

BULB..... ...... 4" × 2"

Whole Length of Girder.	Bearing on Wall at Each End.	Span.	Size of Rod.	Weight of Rod.	Weight of Casting.	Weight of Brick Work.
16 feet.	Say 1 foot 6 inches.	13 feet.	2½ inches Round.	293 lbs.	2,500 lbs.	Say 48 Net Tons.
17 "	" 1 " 6 "	14 "	2½ "	310 "	2,640 "	" 52 "
18 "	" 1 " 6 "	15 "	2½ "	318 "	2,780 "	" 56 "
19 "	" 1 " 6 "	16 "	2¾ "	400 "	2,920 "	" 60 "
20 "	" 1 " 6 "	17 "	2¾ "	420 "	3,060 "	" 64 "
21 "	" 1 " 10 "	17 " 4 inches.	2¾ "	440 "	4,100 "	" 65 "
22 "	" 1 " 10 "	18 " 4 "	3 "	552 "	4,240 "	" 69 "
23 "	" 1 " 10 "	19 " 4 "	3 "	576 "	4,380 "	" 73 "
24 "	" 1 " 10 "	20 " 4 "	3¼ "	700 "	4,520 "	" 77 "
25 "	" 1 " 10 "	21 " 4 "	3¼ "	728 "	4,660 "	" 81 "
26 "	" 2 feet.	22 "	3½ "	864 "	4,800 "	" 82 "
27 "	" 2 "	23 "	3¾ "	1036 "	4,940 "	" 86 "
28 "	" 2 "	24 "	Two 2½ "	950 "	5,180 "	" 90 "
29 "	" 2 "	25 "	Two 2½ "	980 "	5,320 "	" 94 "
30 "	" 2 "	26 "	Two 2¾ "	1240 "	5,460 "	" 98 "

☞ NOTE.—When Iron Columns support the Girders, the bearings will be less.

ARCH GIRDER—EXAMPLE OF COST.

Suitable to sustain four stories of 12 inch brick wall. Length, 25 feet.

Average cost of pattern, flasks, etc.		$2 50
Moulding.—2 moulders, 1 day each	$3 00—$6 00	
2 helpers, 1 day each	1 50— 3 00	
		9 00
Cores		1 00
Facing		1 00
Chaplets, etc.		40
Cleaning		50
Chipping		2 00
Labor, bringing in flasks, taking out casting, etc.		1 00
Sundries		1 00
Painting		1 25
Cartage		3 00
Losage, 1 girder in 10		1 84
Weight, 2,700 lbs. at c. 3.27		88 29
Wrought-iron tension rod, 2¼ inches diameter, including forging, fitting, etc., 520 lbs. at 7c		36 40
Cost		$149 18

Add 20 per cent. profit for selling price.

LINTEL AND CORNICE COURSE—EXAMPLE OF COST.

such as are generally used above first story columns.

Lintel—Weight, say 100 lbs. to foot, at c. 3.27	$3 27	
Moulding, etc.	2 00	
		$5 27

Amount brought forward......................................		$5 27
Cornice—Weight, 45 lbs. to foot, at c. 3.27...................	$1 47	
Moulding, etc......................................	60	
Fitting up at shop..............................	2 00	
Putting up at building...........................	50	
Screws, bolts, files, etc..........................	30	
		4 87
Painting		40
Cartage		20
Sundries...		25
Cost per foot.......................................		$10 99

Add 20 per cent. profit for selling price.

WINDOW LINTEL—EXAMPLE OF COST.

Average cost of pattern..	$0 15
Moulding...	1 20
Facing...	08
Cleaning...	12
Chipping...	15
Labor..	05
Painting...	25
Cartage..	20
Sundries...	10
Losage, 1 lintel in 10..	15
Weight, 90 lbs. at c. 3.27......................................	2 94
Cost..	$5 39

Add 25 per cent. profit for selling price.

WINDOW SILL—EXAMPLE OF COST.

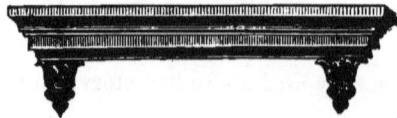

Average cost of pattern..	$0 15
Moulding...	85
Facing................	06

Amount brought forward...............................	$1	06
Cleaning..		12
Chipping..		15
Labor..		05
Painting..		25
Cartage..		15
Sundries..		10
Losage, 1 sill in 10................................		12
Weight, 60 lbs. at c. 3.27..........................	1	96
Cost...	$3	96

Add 25 per cent. profit for selling price.

WROUGHT-IRON PLATE GIRDER—EXAMPLE OF COST.

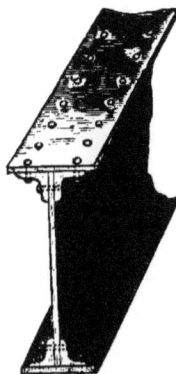

Web, 20″ × $\frac{5}{16}$″.
Top plate, 10″ × $\frac{5}{8}$″.
Bottom plate, 8″ × $\frac{3}{4}$″.
Top angles, 4″ × 4″ × $\frac{7}{16}$″.
Bottom angles, 3$\frac{1}{2}$″ × 3$\frac{1}{2}$″ × $\frac{5}{8}$″.

Length, 30 feet.

Weight, say 2,600 lbs. ; average 3$\frac{1}{2}$c.............................	$91	00
Making—four hours forge and three helpers, $1.30 .... $5 20		
" ten hours forge and three helpers, 71c............. 7 10		
" twenty hours finisher and four helpers, $1.40...... 28 00		
" five hours finisher and three helpers, punching, $1.15 5 75		
" five hours finisher and one helper, 65c............. 3 25		
	49	30
Use of punch, shears, etc...	10	00
Rivets, 120 at 7c..	8	40
Painting	2	00
Cartage and handling..	4	00
Cost..	$164	70

Add 25 per cent. profit for selling price.

RAILING—EXAMPLE OF COST.

Cost of one panel of railing, six feet in length.

Forging... $1 00
Finishing in shop.. 1 50
Lead—2½ lbs. at 8c.. 20
Files, chisels, etc... 20
Painting... 25
Labor, cartage, etc.. 25
Putting up at building... 1 50
Sundries... 50
Weight : Wrought Iron—
    Bottom rail, 1¼ × ½, 6 feet.
    Top rail, 1¼ × ⅝, 6 feet....
    Brace, ¾ × ⅝, 3 feet....... } 35 lbs. at c. 3.22................. 1 03
    Post, ⅞ × ⅞, 3 feet........
    $65 per ton—[see table].
Cast iron—Hand rail........20 lbs.
    Railing castings..60 lbs.
                           —————
                    80 lbs. at c. 3.27.............. 2 61
Pig iron—$30 per ton [see table].
Moulding and other costs in foundry........................ 1 60
                                           ————— 4 21
                                           ═════
   Cost of 6 feet.. $10 64
   Add 25 per cent. profit for selling price.

NEWAL POST—EXAMPLE OF COST.

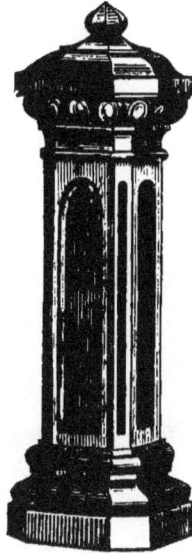

Weight, 115 lbs. at c. 3.27	$3 76
Moulding, and other costs in foundry, etc.	2 30
Finishing	3 50
Screws, files, etc	60
Painting	40
Labor	20
Cartage	30
Cost	$11 06

Add 25 per cent. profit for selling price.

OAT MANGER—EXAMPLE OF COST.

Size inside, 22½ inches by 14¾ inches by 9½ inches deep.

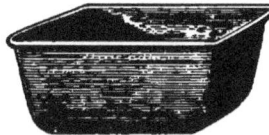

Weight, 60 lbs. at c. 3.27	$1 96
Moulding	10
Facing, moulding sand, handling, etc.	60

Amount brought forward.....................................	$2	66
Cleaning, chipping, files, etc................................		12
Cartage..		08
Losage		10
Cost...	$2	96

Add 33⅓ per cent. profit for selling price.

IRON SHUTTERS—EXAMPLE OF COST.

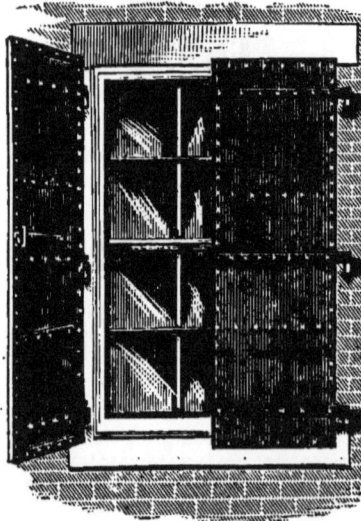

Made in two folds, and hung to eyes built in the wall.

Size, 4 feet wide by 6 feet high.
Frames, 1¼″ × ¼″, covered with No. 16 sheet iron.

Weight.	Lbs.
4 uprights, 6′ each = 24 feet of 1¼ × ¼.............................	60
4 crosses, 4′ " 16 feet of 1¼ × ¼	40
2 hinges, 4′ " 8 feet of 1¼ × ¼·..............................	20
1 striking bar, 5′ of 2 × ¼	16
Latches, rings, etc..	3
Rivets...	5
Wastage, 10 per cent	15
	159

Bar iron ($65 per ton) 159 lbs. at c. 3.22 .... $5 13
Sheet iron, No. 16, 4' 3" × 6' 2", including laps......... 69
Wastage, 10 per cent............................... 7

76 at 5½c..   4 18
Blacksmith and one helper will forge three pair a day, $7.11, is..... 2 37
Finisher and one helper, with shop expenses on same, making, say.. 3 00
Hanging : finisher and one helper will hang six pair a day, say...... 1 00
Cartage (eight pair to a load) and handling, say.................... 50
Painting... 60
Sundries... 50

Cost.. $17 27

Cost per square foot, superficial, 72c.

Add 25 per cent. profit for selling price.

ROOF CRESTING—EXAMPLE OF COST.

Average weight per foot, including two finials to each 25 feet, 10 lbs.
at c. 3.27 .... $0 33
Moulding and other cost in foundry............................... 30
Cleaning, chipping, etc... 05
Fitting up in shop.. 10
Screws, files, etc.. 03
Painting, cartage, sundries, etc.................................. 08
Putting up at building.. 15

Cost per foot.. $1 04
Add 33⅓ per cent. profit for selling price.

4

GRATINGS.

Example of cost of wrought iron gratings 2′ 8″ to 4′ 0′ out, 2 inches centres.

Filling in bars, $1\frac{1}{4} \times \frac{1}{4}$
Front frame bars, $2 \times \frac{1}{4}$ } Not including platforms or doors.
Back frame bars, $3 \times \frac{1}{4}$
On twenty-five feet run.

Finisher and helper, with punching machine, $1\frac{3}{10}$ days at $8..........	$10 40
" " " " $1\frac{1}{2}$ days at $6..........	9 00
Forge, straightening bars and cutting off same, 11 hours at 71c.......	7 81
Painting, 6 hours, and paint 50c......	3 00
Handling...	2 00
Cartage, 25 feet...	2 00
Drilling, etc., for thimbles................................	4 00
Putting down at building: Finisher and helper, $2\frac{1}{2}$ days at $6.50....	16 25

Cost of workmanship on 25 feet.............................. $54 46
Or, $2.14 per lineal foot.

Cost of iron, say, $65 per ton (as per table), is................ 3.22c. per lb.
Wastage, 10 per cent..................................... 32c.

3½c. per lb.

A grating of $1\frac{1}{4} \times \frac{1}{4}$ bars 2 inch centres, 4′ 0′ out, weights $71\frac{5}{10}$ lb. to
foot at 3½c $2 50
Cost of workmanship per foot................................. 2 14

Cost... $4 64
Or, 6½c. per lb.

Add additional for cast iron platforms.   Add additional for grating to raise up, or doors.

TABLE OF WEIGHTS, PER LINEAL FOOT, OF WROUGHT IRON GRATINGS.

(STRAIGHT, NOT KNEE'D.)

1¼ × ⅜ filling in bars—2 inches from centres—6 bars to a foot run.
2¼ × ½ back bar.
1¾ × ½ front bar.
Lead and thimbles, 2 4/10 lbs. to lineal foot.

Out from building.	3'.0"	3'.2"	3'.4"	3'.6"	3'.8"	3'.10"	4'.0"	4'.2"	4'.4"
lbs. per lineal foot.	38 1/10	39 7/10	41 3/10	42 9/10	44 5/10	46 1/10	47 7/10	40 3/10	50 9/10

1¼ × ⅜ filling in bars—1¾ inches from centres—6½ bars to a foot run.
2¼ × ½ back bar.
1¾ × ½ front bar.
Lead and thimbles, 2 4/10 lbs. to lineal foot.

Out from building.	2'.6"	2'.8"	2'.10"	3'.0"	3'.2"	3'.4"	3'.6"	3'.8"	3'.10"	4'.0"	4'.2"	4'.4"	4'.6"
lbs. per lineal foot.	36.7	38 4/10	40 1/10	42 7/10	44	45 8/10	47 1/10	49 4/10	51 1/10	53 1/10	55	56 8/10	58 6/10

TABLE OF WEIGHTS, PER LINEAL FOOT, OF WROUGHT IRON GRATINGS.

$1\frac{1}{4} \times \frac{5}{8}$ filling in bars—2 inches from centres—6 bars to a foot run.
$2\frac{1}{4} \times \frac{3}{8}$ back bar.
$2 \times \frac{3}{8}$ front bar.
Lead and thimbles, $2\frac{5}{10}$ lbs. to lineal foot.

Out from building.	2'.8"	2'.10"	3'.0"	3'.2"	3'.4"	3'.6"	3'.8"	3'.10"	4'.0"	4'.2"	4'.4"	4'.6"	4'.8"	4'.10"	5'.0"
lbs. per lineal foot.	$40\frac{4}{10}$	$42\frac{4}{10}$	$44\frac{4}{10}$	$46\frac{2}{10}$	$48\frac{1}{10}$	50	$51\frac{9}{10}$	$53\frac{8}{10}$	$55\frac{7}{10}$	$57\frac{5}{10}$	$59\frac{4}{10}$	$61\frac{2}{10}$	$63\frac{2}{10}$	$65\frac{2}{10}$	$67\frac{1}{10}$

$1\frac{3}{8} \times \frac{5}{8}$ filling in bars—$1\frac{1}{4}$ inches from centres—$6\frac{7}{10}$ bars to a foot run.
$2\frac{1}{4} \times \frac{3}{8}$ back bar.
$2 \times \frac{3}{8}$ front bar.
Lead and thimbles, $2\frac{5}{10}$ lbs. to a lineal foot.

TABLE OF WEIGHTS, PER LINEAL FOOT, OF WROUGHT IRON GRATINGS.

2'.6"	2'.8"	2'.10"	3'.0"	3'.2"	3'.4"	3'.6"	3'.8"	3'.10"	4'.0"	4'.2"	4'.4"	4'.6"	4'.8"	4'.10"	5'.0"	5'.2"	5'.4"	5'.6"	Out from building.
49	$51\frac{5}{10}$	54	$56\frac{5}{10}$	59	$61\frac{5}{10}$	64	$66\frac{5}{10}$	69	$71\frac{5}{10}$	74	$76\frac{5}{10}$	79	$81\frac{5}{10}$	84	$86\frac{5}{10}$	89	$91\frac{5}{10}$	94	lbs. per lineal foot.

$1\frac{1}{4} \times \frac{1}{2}$ filling in bars—2 inches from centres—6 bars to a foot run.
3 × ½ back bar.
2 × ½ front bar.
Lead and thimbles, 3 lbs. to lineal foot.

$1\frac{1}{4} \times \frac{1}{2}$ filling in bars—1¼ inches from centres—6¾ bars to a foot run.
3 × ½ back bar.
2 × ½ front bar.
Lead and thimbles, 3 lbs. to lineal foot.

TABLE OF WEIGHTS, PER LINEAL FOOT, OF WROUGHT IRON GRATINGS.

1¾ × ⅜ filling in bars—2 inches from centres—6 bars to a foot run.
3 × ½ back bar.
2 × ½ front bar.
Lead and thimbles, 3 lbs. to lineal foot.

Out from building.	2'.8"	2'.10"	3'.0"	3'.2"	3'.4"	3'.6"	3'.8"	3'.10"	4'.0"	4'.2"	4'.4"	4'.6"	4'.8"	4'.10"	5'.0"	5'.2"	5'.4"	5'.6"
lbs. per lineal foot.	$46\frac{7}{10}$	$48\frac{9}{10}$	$51\frac{1}{10}$	$53\frac{3}{10}$	$55\frac{5}{10}$	$57\frac{7}{10}$	$59\frac{9}{10}$	$62\frac{1}{10}$	$64\frac{3}{10}$	$66\frac{6}{10}$	$68\frac{7}{10}$	$70\frac{9}{10}$	$73\frac{1}{10}$	$75\frac{3}{10}$	$77\frac{5}{10}$	$79\frac{7}{10}$	$81\frac{9}{10}$	$84\frac{1}{10}$

1¾ × ⅜ filling in bars—1¾ inches from centres—6⅞ bars to a foot run.
3 × ½ back bar.
2 × ½ front bar.
Lead and thimbles, 3 lbs. to lineal foot.

TABLE OF WEIGHTS, PER LINEAL FOOT, OF WROUGHT IRON GRATINGS.

$1\frac{3}{4} \times \frac{1}{8}$ filling in bars—2 inches from centres—6 bars to a foot run.
$3 \times \frac{1}{4}$ back bar.
$2 \times \frac{3}{8}$ front bar.
Lead and thimbles, $3\frac{1}{16}$ lbs. to foot run.

Out from building.																	
2'.8"	2'.10"	3'.0"	3'.2"	3'.4"	3'.6"	3'.8"	3'.10"	4'.0"	4'.2"	4'.4"	4'.6"	4'.8"	4'.10"	5'.0"	5'.2"	5'.4"	5'.6"
lbs. per lineal foot																	
60	63	66	69	72	75	78	81	84	87	90	93	96	99	102	105	108	111

$1\frac{3}{4} \times \frac{1}{2}$ filling in bars—$1\frac{3}{4}$ inches from centres—$6\frac{2}{3}$ bars to a foot run.
$3 \times \frac{1}{2}$ back bar.
$2 \times \frac{3}{8}$ front bar.
Lead and thimbles, $3\frac{1}{16}$ lbs. to lineal foot.

Out from building.																	
2'.8"	2'.10"	3'.0"	3'.2"	3'.4"	3'.0"	3'.8"	3'.10"	4'.0"	4'.2"	4'.4"	4'.6"	4'.8"	4'.10"	5'.0"	5'.2"	5'.4"	5'.6"
lbs. per lineal foot.																	
$66\frac{3}{16}$	$70\frac{3}{16}$	$73\frac{8}{16}$	$77\frac{1}{16}$	$80\frac{6}{16}$	84	$87\frac{4}{16}$	$90\frac{3}{16}$	$94\frac{3}{16}$	$97\frac{7}{16}$	$101\frac{1}{16}$	$104\frac{9}{16}$	108	$111\frac{1}{16}$	$114\frac{8}{16}$	$118\frac{7}{16}$	$121\frac{7}{16}$	$125\frac{1}{16}$

TABLE OF WEIGHTS, PER LINEAL FOOT, OF WROUGHT IRON GRATINGS.

2 × ½ filling in bars—2 inches from centres—6 bars to a foot run.
3 × ⅝ back bar.
2½ × ½ front bar.
Lead and thimbles, 3 6/10 lbs. to lineal foot.

Out from building.																		
3'.0"	3'.2"	3'.4"	3'.6"	3'.8"	3'.10"	4'.0"	4'.2"	4'.4"	4'.6"	4'.8"	4'.10"	5'.0"	5'.2"	5'.4"	5'.6"	5'.8"	5'.10"	6'.0"
75 4/10	78 8/10	82	85 4/10	88 8/10	92 2/10	95 6/10	99	102 4/10	105 5/10	109 1/10	112 6/10	116	119 4/10	122 8/10	126 2/10	129 6/10	133	136 4/10

lbs. per lineal foot.

2 × ½ filling in bars—1¾ inches from centres—6⁹ bars to a foot run.
3 × ⅝ back bar.
2½ × ½ front bar.
Lead and thimbles, 3 6/10 lbs. to lineal foot.

The foregoing illustrations could be carried to the full extent of showing the cost of every article that pertains to the business. A sufficient number has been given to enable any foundryman to adapt these principles to his own particular class of work, based on the business expenses under which he rests.

The prices of architectural castings do not materially fluctuate with the price of pig iron. The cost of the iron which enters into many of the finished articles is not twenty-five per cent. The principal item is labor. The cost of the labor employed is probably thirty to fifty per cent. greater than in 1860 –'61. This is due not only to the increase in wages, but to the greater care with which work is done. The number of parts and the difficulty of casting them are increasing every year, and more skilled labor is required, in proportion to the amount of iron cast, in the work of fitting up. Of still greater importance, as affecting the cost of castings, is the large amount of capital locked up in patterns, flasks, machinery, buildings, etc., the value of which shrinks every year. Large capital has to be employed, and the proprietors have to work harder than almost any other class of manufacturers. In fact they do double work as manufacturers and as contractors. Patterns are increasing in variety and extent, demanding a continual outlay of money. There is a growing discrimination between the true and the false in this branch of productive industry. A higher, order of taste is being developed, and the tendency is toward more perfectly finished and more artistically ornamented work. This is an encouraging fact for the future of the business, and though it involves increased expenditure, it is one which manufacturers must recognize. The enterprising manufacturer who will meet the popular demand and give artistic excellence, even to the smallest detail, will not lack for patronage.

The cost of most, if not all, of the articles given in illustration may appear excessive. It must be remembered, however, that there are but few of a kind to be made at a time, and at considerable intervals of time apart. To get out the patterns and flasks and shifting of various articles, all takes time, which must be considered. The moulding, time and expenses connected therewith, and risk of losing the casting, is greater when making up a small number than they are in making up a large number. So in the delivery by cartage: a small number of castings and light weight have frequently to be taken for a load. No one need be told that the cost is much more, proportionately, in making two or three castings of a kind than it is in making two or three hundred. It would be a waste of time to recite why the cost is proportionately less on a greater than on a lesser number; it is self-evident to every foundryman.

This printed information and guide is entirely in the interest of the producer. It is to enable him to fully cover the cost of every article, and not to make a profit on one article and a loss on another. Everything should be taken out of the realm of guess-work and brought down to hard facts. If errors are to be made at all, they need to be made on the winning side. The costs had better be calculated excessively than not enough. It is quite probable that after a foundryman arranges a complete line of costs of the various articles he manufactures, on the system here laid out, that the result will be a curtailment of his business by reason of being unable to compete in prices with his neighbors. Such a man need have no regrets. Let him confine himself to such articles as do pay, or raise the standard of his work so as to command a superior price in market. How many men at the end of a year, after doing a large business, are unable to account to themselves for not making money? They cannot discover where any material savings could have been made or greater economy practiced, and yet there is nothing to show for a whole year's hard labor and anxiety. The fault lay in the fact that a considerable portion of their manufactures were made at a loss. The thousand and one small items had not been considered, and a system of self-deception had been continuously practiced, bringing, in the end, disappointment and discouragement. The small items make up a gross sum which is truly astonishing. A smaller business may be done, but it will be a profitable one, and will steer clear of bankruptcy. The making up of detailed tables of costs are generally considered disagreeable duties, and put off as long as possible. They are necessary to success, and if a manufacturer studies his best interest, he will not only make them complete, but often revise them.

In no department should anything be left to guess-work. In taking off quantities from plans, etc., for the purpose of making proposals, it should be done in such a detailed manner as to be readily referred to and compared with the executed work.

In the foundry, a careful and experienced man should act as foreman. Practical ability in turning out good castings is the one great requisite in such a man, and not one full of scientific theories. More money will be made or lost in the foundry than in any other department. The making of unnecessary flasks must be guarded against—those on hand used as much as possible; the stock of weights, arbors, etc., kept as low as the limits of work will allow; wages seen to that nothing above market rates is paid, and the work properly sorted—the common castings to the cheap grade of moulders, the better qualities to the higher grade of moulders. The melting must be looked after to see that the mixtures of iron are properly made, and the cupola charged without waste of material. Economy everywhere must be enforced.

In the pattern shop, a foreman of experience, good judgment and exceedingly careful and correct must be selected. Moulding is rendered difficult or simple as the patterns are made. The patterns for building work are rarely intricate, and the shrinkage of iron and the contraction of castings in cooling are governed by very simple laws.

Between the draughtsman in making working drawings and the foreman of pattern-makers and the foreman of moulders, perfect accord should reign. It is not always possible to design the casting with equal masses of metal throughout, and then the responsibility will devolve upon the founder, who must, by accelerating cooling of parts by early uncovering, or by retarded cooling of other parts, produce a simultaneous rate of cooling throughout the casting. Great care must be exercised in making patterns to secure a proper distribution of metal. This arises from the fact that in cooling the thinnest parts of the casting becomes quite cool, while the heavier parts are yet red hot. The part which has cooled first having contracted and set, while the other portion is yet soft, the result is that the casting pulls apart in the mould, or is left with a strain and tension which, upon being subjected to a sudden jar, or even to the influences of the weather in expanding or contracting the iron, will produce after breakage.

In the finishing department, the foreman must have a thorough practical knowledge of his branch of work, and ability to control the men under him and get out of them all the work possible. And he must not only have the drive and snap in him, but the workmanship of his men must be good, as well as expeditiously done.

Over all, the care and watchfulness of the manager must be omnipresent. Waste must be prevented, each department made to work systematically and harmoniously with every other, surplus men cut off, and the pay-rolls kept within the closest bounds. Supplies must be bought at the lowest ruling rates, and every item in liberal quantities. The shop must be kept well supplied with work. If good results are to be obtained from journeymen, they must have confidence that there is a full quota of work ahead. Otherwise they will nurse their task in order not to do themselves out of a job. The work must be regular, and not spasmodic. Men work with a will and do their best in busy times, and the reverse of this in dull times. The beginning and the end of the business rests on the manager—on his industry, patience, skill and experience.

The foundry business is peculiar in one respect. The manager has continually to overcome a tendency to name lower rates per pound in taking orders than the facts of the real cost of production warrants. The business is carried on for the purpose of making money, and that aim needs constantly to be enforced by thorough and systematic arrangement of and reference to table of costs.

The field is broad enough without calling forth an unhealthy competition. Frequent and friendly intercourse between those engaged in the same pursuits, and comparisons of opinions and experiences, contribute to the common good. What effects the prosperity of one affects more or less the prosperity of all. It is certainly desirable to know positively what products cost, and to establish prices which allow fair profits. Those engaged in the manufacture of iron work for buildings need to take a broader comprehension of their business. The magnificent proportions which the manufacture of this class of iron work is to assume in the future can scarcely be realized.

SPECIFICATION

OF

IRON WORK AND MATERIALS AND LABOR

required to build and complete a.................. to be erected on lot No...., ... Street, for

Mr. ....,

Owner,

in accordance with this specification and the accompanying plans, elevation, section, and working drawings made by

....

Architect.

DIMENSIONS.—The size of the building, heights of stories and other dimensions to be obtained from the drawings and the figures thereon.

SCRAP.—Take down and clear away all the old iron from present buildings, and allow the value of same in making the estimate.

FRONT.—The front of the building, from the foundation up to the roof cornice, will be made of cast iron, as shown on the drawings, including all posts, antaes, columns, piers, jambs, reveals, arches, facias, cornices, capitals, bases, water tables, sills, panels, and other architectural features. The posts, columns, etc., upon which dependence is placed for stability, will average three-quarters of an inch in thickness. The remainder of the work will be cast of sufficient thickness to retain their shape, none being less than a quarter of an inch in thickness. And the whole securely bolted, and properly put together in their several parts. The castings to be smooth, straight, sharp and clean. The ends of all columns to be turned off true and even in a lathe. Columns to have ears cast on at top and bottom, and bolted together with a plate one inch thick intervening. Longitudinal ties of $2 \times \frac{1}{2}$ inch wrought iron will connect the columns of each story, with an eye formed on each end of the tie bar to allow the passing through of the bolts which connect the columns together by the ears. Bolts three-quarters of an inch in diameter and nuts.

The basement posts will rest on plates one-and-a-quarter inches thick, planed on top, and to be four inches on all sides larger than the posts which rest on them. The plates on top of basement posts will be made with nosings on the front, and sockets on the sides to receive bars for the support of the door sills.

Cast or bolt on necessary brackets, etc., of suitable shape and thickness.

End columns to be each secured with three wrought iron holdfasts, and each middle antae secured by heavy wrought iron anchors extending back about five feet, and well fastened to the floor timbers.

The entire front to be done in the most substantial and workmanlike manner.

Each casting thoroughly coated with paint on all surfaces, including bolt holes, to be painted before using. All bolts to be dipped in paint before being used, and all screws, rivets, etc., to be treated in this way as well ; carefully scrape away all burs, etc., after the drilling of holes.

The joints made flush and true, and water-tight.

Notes.—For a corner building : " The antaes of upper story to continue up to the height necessary to support roof timbers."

For a double store : " A bracket cast or bolted on to the back of the centre front antae on each story to take the end of wooden girders. And these antaes to be each one-half inch thicker than the others are called for in the respective stories."

For a corner building where the floor timbers rest on the return front : " All the front antaes on the side street shall have brackets cast on their sides to receive seven-inch wrought iron beams to support the wooden floor timbers.

These beams shall be made to fit snugly between the antaes, and be fastened with angle pieces bolted through the web of the beam."

If only the first story is iron—that is a brick or stone front used above—then specify: "Over the first story columns a box lintel course will be placed, made, say, twenty inches high on face, say twenty-two inches on bed, and say, twelve inches on back, and to average one-and-a-quarter inches thick. The lintels jointed over the centre of columns, turned off on ends in lathe true and even, and bolted together through brackets cast in lintels, with two three-quarter inch bolts and nuts to each joint. Cornice over lintel course to be thoroughly put up, bolted and bracketed to lintels."

Rolling Shutters.—The front openings of first story to be fitted with revolving iron shutters of approved make, with all shafts, gearings, cranks, chains, counterweights, guides, grooves and all other necessary fixtures complete, put up in the best manner, and left in perfect working order.

The wooden doors beneath the rolling shutters to be covered with sheet iron on the face and edges, properly screwed on. In the panels plant on cast iron mouldings.

The entrance to loft will be fitted with dwarf doors made in four folds, frames of $1\frac{1}{2} \times \frac{5}{8}$ wrought iron, covered with No. 16 sheet iron, panelled and moulded and hung to the iron columns, and furnished with strong bolts and a $15 lock and two keys.

Illuminated Platform.—The steps, risers, platforms and door sills covering front area, shall be of approved make of illuminating tile, consisting of cast iron plates seven-eighths inch thick with knobs on top, and thirty-three double convex lenses set in cement to each square foot of tile. The tile set in cast iron frames, and the latter supported on strong cast iron bearers, beaded on the lower edge. Covering the end walls will be used an iron tile in imitation of the glass. The ends of the walls are to be covered with iron plates extending down below the walk. The exterior surface of frames to be grooved, and edges to be trimmed with nosings. The platform and steps to have proper pitch to the street of about three-quarters of an inch to the foot. That part of the frame which receives the wrought iron doors over elevator will be set on a greater incline, as shown on the drawings. Let into granite a cast iron shoe, from which the first riser will start. The door sills will each be supported on two $3 \times \frac{3}{4}$ wrought iron bars, the whole made complete and put down perfectly water-tight, and kept so for one year from completion of the building.

The elevator doors to be hung and secured to the cast iron frame, to be made in two folds, with frames of $2 \times \frac{1}{2}$ inch wrought iron and covered with No. 12 sheet iron. Have proper padlock fastenings to secure the doors when shut, and have guard bar of seven-eighth inch round iron for protection when doors are open. Eyes to receive this guard bar to be riveted on at both front and back of doors.

Note.—Sometimes the door sills are checquered plates. Sometimes the risers are plain iron. Sometimes the first riser only is plain, the others illuminated. Sometimes checquered plates are introduced in front of the basement columns, in order to reduce the amount of illuminated work, and so cheapen the cost. Sometimes a cheaper platform is required. Then specify: "Cast iron tile glazed with $5\frac{1}{4} \times 5\frac{1}{4} \times \frac{1}{4}$ glass, set in with putty cement, and made water-tight."

VAULT GIRDER AND COLUMN.—Furnish and set in vault, for support of granite sidewalk, a cast iron girder averaging one-and-a-quarter inches thick, made fifteen inches high on the back flange, six inches on the front flange, and eight inches on the bed; the column underneath the girder to be seven inches diameter and one inch thick, and to have bottom plate $12 \times 12 \times 1\frac{1}{2}$ thick, with ring cast on to receive and hold column; the top plate to be cast on the column, the plate to have a moulded turn-up piece on top to prevent the column from being shifted away from the girder.

Note.—If brick arches and stone flags are used instead of granite, then there requires to be specified: "Vault Beams—to be in number as shown on

4

plan, of cast iron ten inches on the bed, with centre flange ten inches high in middle and four inches on ends. Thickness to average one inch. Beams to be made to hook over and lay into girder so that the bed of the beams will be on the same level as the bed of the girder." Wrought iron beams extending from the front basement columns to the street walls, with headers at the area line, are often used.

INTERIOR COLUMNS.—Inside columns supporting wood girders will be as follows : Basement and first story, twelve inches diameter, one-and-a-quarter inches thick. Second story, ten inches diameter, one-and-a-quarter inches thick. All the foregoing to be deep fluted, and have full Corinthian capitals and Attic bases, with round shell plinths four inches high. Columns in third story, nine inches diameter, one-and-a-quarter inches thick. In fourth story, eight inches diameter and one inch thick. In fifth story, seven inches diameter and one inch thick. And these all to have plain shafts, loose Tuscan capitals with egg and dart bed moulding, and moulded bases.

The columns are all to be turned off true and even on ends. Those for upper stories are to be made with dowel ends to pass through girders. And all to have top plates eighteen inches long by the width of girders, and to be one-and-a-half inches thick.

Under the basement columns place cast iron bed plates $18 \times 18 \times 1\frac{1}{2}$ inches planed on top, and bedded level underneath with a small quantity of cement.

The interior columns will be delivered at the building to the framer, who will set them up in place.

Note.—The following applies to a double store :

ARCH GIRDER.—Furnish and put up for the support of the rear wall two cast iron arch girders with tension rods. The top flange of girder to be sixteen inches wide and one-and-a-quarter inches thick. Centre web twelve inches high and one inch thick. To have twenty inch bearings at each end and skewbacks thereat. Rise of girder at centre to be two feet six inches. Tension rod three inches diameter of best refined wrought iron, with square heads at ends, adjusted to the girder in the best manner, and shrunk in while hot. These girders to be well bolted together in the centre, where they meet with four one-inch bolts. Also to be properly bolted to the fire-proof column on which they rest.

Note.—Sometimes two rods are used instead of one. Then they are smaller size rods, the two making in section about fifty per cent. more than the single rod. Sometimes *square* rods are used instead of round.

Fire-proof Column.—For the support of the arch girders, provide and set up in place a double fire-proof column made as follows : Outside column, sixteen inches diameter, and average three-quarters inch thick, and to correspond in style with the other first story columns. Inside column, twelve inches diameter and one-and-a-quarter inches thick, plain shaft, with top plate two-and-a-half inches cast on. Bottom plate, $24 \times 24 \times 1\frac{1}{2}$. On the inside column cast on bracket to receive wooden girder. The space between the outer and inner column to be filled in with plaster.

Note.—Sometimes a fire-proof column is used in the basement underneath the first story column, instead of a brick pier. If so, specify this kind of a column to be used both in the basement and first story, and to be strongly bolted together.

Sky-Light.—Curved sky-light over first story extension to be of illuminated tile supported on handsome moulded cast iron ribs. To have cast iron moulded gutter on top of extension wall with flanges to fit over the thickness of wall. The bottom of gutter to have a fall to one end, and to have proper outlet to receive leader.

The leader pipe to be of cast iron four inches diameter, running down on the outside of wall, and connecting with the drain pipe in cellar. The joints of leader properly leaded and caulked and made tight. Put a cast iron strainer over leader in gutter.

The end brick walls coped with cast iron and turned down over outer side four inches.

Cover the outer face of arch girders and the brick-work of the rear wall above up to the under side of the second story window sills with No. 14 sheet iron, riveted to two-inch angle iron at top, the angle iron to be furnished to the mason and to be built in the brick wall by him.

Note.—Sometimes the gutter pitches both ways—from the centre to each end.

If a pitch sky-light is wanted, then specify as follows: Over the rear extension of first story provide a pitched sky-light formed of wrought iron sash bars $\frac{1}{2} \times 2$ inches, placed eight inches apart, with one-and-a-quarter inch oval rebate bar riveted on underneath. Front frame bar, $2\frac{1}{2} \times \frac{1}{2}$; back frame bar, $3 \times \frac{1}{2}$; sky-light glazed with rough plate glass three-eighth inch thick, in sheets as long as the sash bars.

Note.—Gutter, leader-pipe, coping, etc., as before specified. *Sometimes the glass is put in carpenter's specification.

FLOOR-LIGHTS.—Make and put down in first story floor-lights of wrought iron, as per plan. Main bars to be $\frac{3}{4} \times 3$, with $\frac{1}{2} \times 1$ rebates riveted on; cross bars, 1×1, to run through main bars; to have a cast iron border of neat pattern, four inches wide on top, and made with sockets on the sides to receive ends of bars, and to be well fitted around to the floor, and securely screwed down. Glaze with rough plate glass, one inch thick, well bedded in putty.

Note.—Sometimes where a floor-light is very large, nine inch (or other size) wrought iron beams are used to secure rigidity. The glass used for floor-lights is generally one inch thick, generally twelve inches wide, and from thirty to thirty-six inches long. Even size glass should always be used, that is, ten, twelve, fourteen inches and the like wide—not ten-and-a-half, eleven, etc.

REAR OUTSIDE SHUTTERS.—Supply to all the rear openings outside shutters, in two folds to each window, three panels for those seven feet or over in height, and two panels for those under seven feet; made with frames of $\frac{1}{2} \times 1\frac{1}{2}$, and covered with No. 16 sheet iron, riveted to frames with rivets placed about four inches apart, the sheet iron to lap full one inch on the brick-work; to have strong wrought iron strap hinges extending the full width of shutter, and to be well riveted to and through the frames; shutting bars of $\frac{1}{2} \times 1\frac{3}{4}$, built in brick-work; all the shutters to be furnished complete with all required rings, latches, staples, turn buckles, etc.; the shutters of basement, first and second stories to have three-quarter inch square bolts at top and bottom, in addition to the other fastenings; the top bolts to be long enough for convenient reach from the floor below; the bottom bolts to shoot over wrought iron stubs, which must be leaded in the stone window sills; all the shutters to be left in easy working order.

Note.—Double shutters are frequently used; shutters having a double covering of sheet iron, with an air space between.

SHUTTER EYES.—Furnish to the mason to build in the wall cast iron brick-eyes for all the rear window openings; three to each jamb for openings of seven feet or over, and two for each jamb of openings less than seven feet in height.

WINDOW GUARDS.—Front windows at hoistway in the several
stories will have guards extending from sill to soffit of
windows, made of three-quarter inch round bars placed
five inches from centres, with top and bottom rail and two
centre cross bars of $\frac{1}{2} \times 2$ inch.

Note.—For windows on side street or for rear windows in basement or first
story, then specify : Have guards of seven-eighths round wrought iron bars,
sharp pointed at head, or with ornamental cast iron heads, extending up to
within four inches of soffit of window, placed five inches from centres, and
passed through two $\frac{1}{2} \times 2$ inch cross bars leaded in brick work. The bottom
ends of round bars leaded in the stone window sill. Sometimes one window
guard is arranged to open on hinges and secured with strong hasp, staple and
brass padlock.

ANCHORS.—Provide a sufficiency of strong wrought iron anchors, ties, clamps, etc., of every description that may be
required to completely finish the building, including anchors to secure the cast iron front. Beam anchors placed
not over seven feet apart, made of $\frac{5}{8} \times 1\frac{3}{4}$ iron, with one
inch round spear bar twelve inches long ; the end which is
fastened to the beam to be hook-shaped, laying over a
wrought iron spike driven in side of beam.

Note.—If specifying for a double store, then there is required : Girder straps
made of $\frac{1}{4} \times 2\frac{1}{2}$ iron, twenty-four inches long, with wrought iron spikes, two
straps to each joint of girder.
Sometimes star anchors are used on the ends of girders, storehouses and
heavy warehouse buildings requiring them. The star is made of cast iron
from eight inch up to fourteen inch diameter. The shank of $\frac{1}{2} \times 1\frac{1}{4}$ wrought
iron with bolt end and nut, and punched for spikes.

BRIDLE IRONS (or Stirrup Irons, sometimes called).—Furnish to
the carpenter as required : Bridle irons for all headers
and trimmers of stairways, hoistways, floor-lights, fire-
places and flues, made of $\frac{1}{2} \times 3$ best refined wrought iron,
free from flaws and other defects, and furnish wrought
iron spikes, of suitable size, to secure them to the timbers.

BOOK VAULT DOORS.—Outside hanging frames $2 \times \frac{3}{8}$, inside $2 \times \frac{1}{2}$, properly doweled together through the brick wall. The frame of outside door made of $2 \times \frac{1}{2}$ in one fold and covered with one-eighth inch boiler plate. Frame of inside door $1\frac{1}{2} \times \frac{1}{2}$, made in two folds and covered with No. 14 sheet iron and secured with swivel bar; the outside door to be panelled and moulded, and furnished with a $15 lock; the vestibule lined at top, bottom and sides with No. 16 sheet iron.

GIRDERS AND FLOOR BEAMS.—The girders throughout will be formed of rolled wrought iron beams, bolted together with one-inch diameter bolts and nuts passing through cast iron thimble-pieces, of such shape as to hold the beams evenly and closely together. These blocking pieces placed at each end of girders, and intermediately, not exceeding three feet apart.

The girder beams under basement and first story will be *twelve inch heavy, weighing each 180 lbs. to yard.* Those under second and third story floors will be *twelve inch light, weighing each 125 lbs. to yard.* Those under the fourth and fifth story floors and roof tie will be *light ten-and-a-half inch, weighing each 105 lbs. to the yard.*

All these girders are to be bolted and fastened to the columns in the strongest and most approved manner.

The floor beams throughout will be of rolled wrought iron, and placed the distances apart as shown on framing plans.

Those for basement and first story floors will be *twelve inch light, weighing 125 lbs. to yard.* Those for second and third story floors will be *nine inch heavy, weighing ninety lbs. to yard.* And the tiers above, including the roof, will be *nine inch light, weighing seventy lbs. to yard.*

Each bay of beams will have one row of seven-eighths inch diameter bolts and nuts, forming tie rods.

All framing beams, headers and trimmers, will be double, bolted together, and the headers and trimmers framed together and connected with 4 × 4 angle irons and seven-eighths inch bolts and nuts.

On each tier the beam against front and rear walls will be of channel iron, corresponding in height and thickness with the respective floor beams.

Note.—Sometimes a "plate" girder is used made with centre web, say 14 × $1\frac{4}{8}$; top plate, 10 × $\frac{1}{2}$; bottom plate, say 8 × $\frac{5}{8}$; angle iron, $3\frac{1}{2}$ × $3\frac{1}{2}$ × $\frac{1}{2}$, with cast iron shoes bolted on side to receive the floor beams.

GALVANIZED CORNICE.—The main cornice, *balustrade, dormer windows, angle ribs, etc.,* of roof *and pavilions,* to be of No. 24 galvanized iron, with zinc ornaments. The joints in the galvanized iron riveted together as well as soldered. The cornice to be straight and true and perfectly tight, and well and securely sustained and retained by strong wrought iron straps.

CRESTING.—Put up a cast iron crest railing of approved pattern, with end finials, screwed down to roofing plank with screws bedded in white lead, and the cresting strongly braced and made secure.

GRATINGS.—The basement areas are to be enclosed with wrought iron bar gratings, made of $\frac{1}{2}$ × 2 frame bars and $\frac{1}{2}$ × $1\frac{3}{4}$ inch filling in bars, placed one-and-three-quarter inch from centres. Said gratings to be stiffened with rods and thimbles, and to be well secured to the stone coping and to the brick piers.

In front of the doorways have perforated cast iron walking plates.

Note.—Sometimes "knee" gratings are required—that is, forming steps and risers.

Sometimes area light holes have to be covered with gratings; these (or one of them) are usually made to open on hinges, and secured with chain, staple and padlock.

ELEVATOR.—Put up in area, from sidewalk to lower story floor, a platform elevator, with iron frame hoisting apparatus, guides, chains, cross-heads, heavy wooden platform, strongly braced, all made complete and done in a substantial manner.

SCREEN WORK.—At the foot of stairs in first story cover the entire sash partition, from floor to ceiling, with wrought iron lattice work of a neat selected pattern, made of $\frac{1}{8}'' \times \frac{3}{4}''$ scroll iron, well riveted, with frames of $1'' \times 1''$ properly secured to the woodwork. Put door in same, with brass padlock and fastenings.

Cover glass in partitions around hatchway in basement and first, second, third and fourth stories.

Lattice guards under each store window of the front, securely fastened in.

SADDLES.—Make and screw down cast iron door saddles to all first story doors, from patterns to be furnished by the carpenter, perforated for bolts of doors to shoot in.

PLATES FOR WOODEN COLUMNS.—Furnish top and bottom plates for all the wooden columns, of cast iron, sixteen inches long by the width of girders, under and on top of which they may be used. All to be one-and-a-half inches thick, each to have a raised oval projection on one surface to take the dowel, and on the other surface a raised moulding to hold the column, round or square, as the case may be.

DOWELS.—Furnish cast iron dowels, oval shaped, 5×10 inches diameter, three-quarter inch thick, and in length one-quarter of an inch less than the depth of the wooden girders through which they pass.

Note.—Sometimes three wrought iron dowel pins are used, one-and-one-quarter inches diameter.

Sometimes cross-shaped cast iron dowels are used.

FLUE DOORS.—Furnish to mason to build in wall, for all the

flues in cellar, doors 12″×12″, hung and latched to wrought iron frames.

FLUE RINGS.—Furnish cast iron flue boxes, and five-and-one-half inch diameter removable ventilating covers to same, for all flues on each story. The boxes to be furnished to mason to build in, and the covers to be set on after the plastering is done.

DRILL FOR CARPENTER.—Do all drilling and tapping required by carpenter to secure his wood work to the iron work, and furnish all screws and bolts required for this purpose. Do all cutting and fitting that may be required in connection with the iron work.

COAL COVER.—Furnish one cover to coal vault, with cast iron neck; the latter let into the granite sidewalk. The cover to be quarter-inch wrought iron plate, studded with rivets on top to prevent slipping, secured with proper galvanized chain, hook and staple.

Note.—Often an ordinary cast iron coal cover, 18′ diameter, with chain and fastenings, will answer.

VENTILATING GRATING.—Furnish and fasten into the rebate, to be cut in granite sidewalk, a cast iron heavy grating for ventilation, 14×14 inches square, and below it have a movable slat register 14×14, hung on hinges, and a sheet iron drawer to catch and hold the dirt falling through the sidewalk grating.

SCUTTLE.—Cover the wooden scuttle door on the underside; also line the inside of scuttle curb with No. 12 sheet iron. At the level of ceiling make and put up a grating door to scuttle, the frame of $\frac{3}{8} \times 1\frac{1}{4}$, the lattice filling of $\frac{3}{16} \times \frac{3}{4}$, properly hinged, and arranged for and supplied with padlock.

Note.—Oftentimes the lining only is required. Sometimes the lattice door, and not the lining.

LADDER TO SCUTTLE.—A wrought iron ladder will be required to scuttle, made of $2 \times \frac{3}{4}$ sides, placed eighteen inches apart, and the rungs to be double and five-eighths of an inch diameter, twelve inches apart, put strongly together. Set in place at an angle a little less than vertical, and securely fastened at top and bottom.

FIRE ESCAPES.—The platforms of all balconies shall not be less than two feet and six inches in width from the face of the wall, and shall take in two windows in length, either on the front or rear of all double tenement buildings to which they are attached; and the balconies on all single tenement houses shall be constructed precisely the same, except that they shall not be less than six feet in length and take in one of the room windows.

The bearing bars or brackets shall be made of bar iron, not less than one-and-one-half by one-half inch in thickness, and the braces not less than three-quarter round iron, well riveted to the bearing bars; the guard rails to be not less than two feet and six inches in height, and the top and bottom of the same not less than one-and-one-half by three-eighths inch wide; the filling in bars may be made of cast or wrought iron, well secured to the top and bottom rails; if of wrought iron, the bars must not be less than five-eighths inches round, and placed not more than twelve inches from centres. The floors of the balconies may be made either of wood or iron; if of iron, the bars not to be less than one-and-one-half by one-half inch, and not more than one-and-one-half inches apart; and if of wood, to be of yellow pine not less than one-and-one-quarter inches thick, a trap door arranged in each. The bearing bars and braces and the top and bottom of the guard rails are to be let into all walls, twelve inches or over in thickness, at least six inches, well fastened, and in all walls less than twelve inches thick, the same to go through the walls, and be well secured on the inside. The ladders to be made of wrought iron, the side bars of $1\frac{1}{2} \times \frac{5}{8}$ inches, the rungs to be five-eighths inches round, and placed not more than fifteen inches apart. The ladder to extend from the ground to the roof. The lower section—the one from the ground to the first balcony—arranged to slide up, and with hook to hold it in place. The upper section to have circled ends, and at a proper distance above the level of roof, so as to form a safety hand-rail. All to be well fastened.

FUEL ROOM.—Line the fuel room with No. 16 sheet iron, properly lapped and well nailed to the wood studding. The inside of the doors and ceiling included, as well as the floor and sides of the room to be lined.

CHIMNEY CAPS.—The chimney caps to be each in one piece of cast iron one-quarter inch in thickness, arranged for flues, and turned down two inches over the brick work, both in the flues and over the extreme projection of the brick work. To be made slightly crowning toward the centre.

CHIMNEY SHAFT.—Cast iron bases, mouldings in the brick panels and moulded tops to the chimney shafts, to be made as shown on drawings, and furnished when required to be built in.

COPING.—Cover the parapet walls with cast iron copings one-quarter inch in thickness, properly crowning on top to shed water both ways, fitted with lap joints, and well put together with screws. To turn down over wall two inches on each side. To be painted two coats metallic paint inside and outside before being put up.

WINDOW LINTELS AND SILLS.—Furnish to the mason when required cast iron moulded ornamental outside window lintels and sills of the designs shown on the elevations. To be about five-sixteenths of an inch in thickness and to have eyes cast on the inside, and furnished with suitable wrought iron retaining anchors. The top shelves to have turn-up lips to prevent the water getting in behind. To be thoroughly coated with metallic paint, two coats inside and outside, before being put up in place.

TRIMMINGS.—The lines of quoin blocks, ashler bands, etc., will be of cast iron, moulded and ornamented as shown, $\frac{5}{16}$' thick, painted two coats inside and out, before being set up. Each casting to have an eye on the inside, and furnished with suitable wrought iron anchors. All fitted and put up straight and true.

COAL VAULT DOOR.—Put up to the coal vault a sheet iron door two feet wide by six feet in height, hung to eyes built in the wall. Frame of door, $2 \times \frac{1}{2}$ wrought iron, covered with No. 14 sheet iron, properly made and hinged. The lower part of the door, two feet in height, made to turn up and properly hinged. To have strong hinges, bolts and fastenings complete. To be panelled and moulded.

PORT-HOLES.—In the parapet or fire walls, which are carried up about five feet from the roof level, place cast iron port-holes ten feet apart. These port-holes will be made in one casting, in shape like an hour-glass, six inches diameter of opening in the centre, and radiating to a larger diameter of twelve inches. In the small openings place panes of mica.

BALCONIES.—Fancy balconies, composed of ornamental cast iron brackets and selected pattern of railings, all properly braced and tied together with wrought iron bars, will be put up where shown on the elevations. The bars on the brackets will be made to go through the brick wall, and have plate washers and nuts on the back. To be arranged to receive wooden floors.

STAIRS.—The main staircase, as shown on plans, will commence at the level of first or principal story and run from thence up to the level of the upper floor. To be of iron, resting on three good and sufficient close stringers of cast iron, one-half inch thick in the thinnest part. The stringers are to be moulded and beaded so as to form a finished appearance underneath, and to have shoulders cast on them to receive treads and risers. The wall string to have a moulded skirting, the skirting carried around the platforms. The risers to be not less than one-quarter of an inch in the thinnest part, panelled inside and out. Steps and platforms one-half inch thick, checquered on top, with fluted margins, and ribbed underneath. All to be well connected with necessary braces, angles, flanges, ties, bolts, etc.

All stairs to be seven inch rise by thirteen inch tread, including the nosing, which overhangs one-and-one-half inches; thus making the going of the steps measure on the strings $7 \times 11\frac{1}{2}$ inches.

Newel posts will be as shown, properly secured at all angles of stairs and landings where laid down on plans. Ornamental balusters of cast iron, bolted to the face of outer string, and have wrought iron top rail of $\frac{1}{2} \times 1\frac{1}{2}$ inch drilled, to fasten the wooden hand-rail. The balusters to run along the stairs and well-holes.

The stringers to be well secured to the brick walls, and all necessary short beams at the platforms to be provided. The joints neatly made, and all put together in a workmanlike manner.

Note.—The strings, or carriages, may be of wrought iron, or a combination of wrought and cast iron. The soffits may be lathed with sheet iron lath. The treads may be of wood, such as oak, yellow pine, etc., or of marble or other stone, and the iron strings properly prepared to receive them. If the stairs are very wide, two intermediate strings may be required in addition to the outer and wall string. Landings may be supported on wrought iron beams. Open-work risers may be used, or risers entirely omitted. A bracket hand-rail along the wall may be used. Railing balusters and hand-rail may be of square or round iron or gas-pipe. The hand-rail may be moulded, of cast iron.

HOOKS.—Wrought iron hooks for clothes-line pulleys, awning hooks, and hooks for looking-glasses, as may be required, of suitable shapes and sizes, will be furnished to the mason to be built in by him.

PAINTING.—Prime all the iron work with one good coat of brown mineral paint in oil before being brought to the building. All cornices, ornaments and fittings must be painted both sides. All bolts, screws, etc., must be dipped in paint before being used. All joints puttied up and made tight.

All the iron work must be painted two good coats, in addition to the priming coat, of the best English or selected American white lead and oil, in such colors as may be required, except the top of area platform, which latter will be painted in two coats black tar paint. The following may be observed, unless otherwise directed : The fronts, interior columns, underside of vault work, underside of skylight, etc., in white; the rolling shutters and outside rear shutters in green.

GENERAL REQUIREMENTS.—All the castings to be of the best quality of cast iron, straight and smooth and well fitted, with neat, close joints. All the materials to be of the best quality of their several kinds. The entire work is to be executed in the best and most workmanlike manner, and in strict conformity with the particulars set forth in this specification and the several drawings. Any work shown on the plans and not included in the specification, or *vice versâ*, or that may be required to finish the job complete, to the true intent and meaning of the plans and specifications, shall be done by the contractor without any extra charge whatever.

All the work is to be so done as to meet the requirements of the building laws as now in force in the district where this building is to be erected. Proper tests and examinations of the iron work shall be made by the contractor, and he shall be held responsible for any and all damages. All columns, lintels, girders, etc., to have the maximum weights they will safely sustain marked or stamped thereon.

All the work is to be done to the entire satisfaction of the owner and of the architect.

The building is to commence on ——— day of ——— next, and to be completed, fit for occupancy, on or before the ——— day of ——— following.

IRON FRONTS.

For building purposes, cast iron possesses unequalled advantages of strength, durability, economy, and adaptability to ornament and decoration. In resisting any kind of strain, it is vastly superior to granite, marble, sandstone, or brick. Practically, cast iron is crushing proof, for a column must be ten miles in height before it will crush itself by its own weight. Unlike wrought iron and steel, it is not subject to rapid oxidation and decay by exposure to the atmosphere ; and whatever tendency it may have in that direction can easily be prevented by a proper coating of paint. No other material is so valuable after it has served its original purpose, as it may be recast into new forms and adapted to new uses.

In business quarters, where blocks of stores are built up solid, where each building nearly covers the full lot, rear almost butting to rear, with window openings generally only at the front and back, light becomes one of the most important requirements. A light edifice of iron may be safely substi-

A REPRESENTATIVE IRON FRONT.

6

tuted for the cumbrous structures of other substances, and ample strength secured without the exclusion of daylight. Iron in this respect presents peculiar fitness.

The introducing manufacturers and architects in iron acted on the self-evident proposition that a multiplicity of ornament and decoration could be executed in iron at an expense not to be named in comparison with that of stone, and literally covered their fronts with useless filagree work. Every column was made fluted or of some intricate pattern, every moulding enriched. The carvings high up in the air, on the fifth story, were the same as those low down on the first—no bolder, and in every case too flat and fine. Instead of seeking for beautiful outlines and proportions, and appropriately embellishing special features to contrast with other portions of the edifice purposely left plain and unpretending, ornateness was made the governing idea, and an extreme elaboration produced, with twistings and contortions of outline, and crowding in of small columns and pilasters, and diminutive friezes and cornices, overlaying everything with so-called ornament. Constructors in iron took advantage of the ability of cast iron to resist compression, and of the tensile power of wrought iron, and in an utilitarian spirit produced spider-like structures, suggesting nothing save economy of space and material. Overloading the surface with poorly executed ornament, gave their structures a flashy and vulgar appearance. These early stages have been passed, and taste and utility now go hand in hand. For a time, the material was judged more from the mistakes of the unskilful than by its capabilities for proper application.

A building should bear the impress of solidity, as though it were indeed a growth of the earth itself, and not of so fragile an appearance that the winds can blow it away. In architecture, the recognition of permanency is one of the true principles of the art. A front must not only be strong enough—it must also possess such an evident reserve of strength, which is the result of obvious abundance. Convenience, permanence and beauty, as well as strength, are the tests of iron work. And constantly large columns are used where smaller ones would answer. A broad play for light and shadow should be carefully studied. Ornamentation should not be made an end, but a mere adjunct. If beautiful outline and proportion be lacking, no amount of meretricious ornamentation can supply the deficiency. Iron affords a cheaper material, a more enduring material, and cleaner and sharper than stone, and it is the best material, all things considered, for the street architecture of our American cities. Whatever moulding is good in stone, for projection or general outline, is also good in iron. If the ancient examples of cornices and capitals, and ornaments generally, which have stood the test of criticism and been judged correct, are deemed best for stone, then they are best for iron also. But correct outlines must be faithfully followed, and can be in the hands of a skilful manufacturer. If error be committed by the unskilful, it no more condemns the material than will the thousands of ludicrous mistakes in wood and stone condemn those materials. The ancients worked in stone, and artistically produced outlines that perhaps never can be rivalled. The principles of architecture, which have endured so long, will remain forever, simply because they embody true taste and common sense, both of which the public have and understand. On the presumption that the public possess no taste, gross incongruities in design are too often put upon the credulity of those who build. Here a great mistake is made. The public eye is a sharp one, and demands to be pleased. Whether there is an educated or a natural taste, there is at least *an opinion* to be gratified ; and in such cases the majority rules, for, though all do not think alike, a vast number may come to one conclusion, and *that* is generally sure to be correct. Iron is the modern building material, dug from the bowels of the earth, smelted and purified by an advanced science, and ready to supplant stone, just as history relates stone supplanted mud in the construction of dwellings for men. Each tells of a growth in knowledge, applying a better material. Long after a stone front has gone to decay and disappeared, the iron will be retained in its original fulness and sharpness in every line. Keep it painted, and after a thousand years of exposure to the wind and weather, an iron front will be as perfect as on the day of erection.

To paint iron costs much less than to paint wood or other materials, on account of its non-absorbing surface. The *interest* on the difference in first cost between a stone and an iron front will easily pay for one coat of paint a year. More than that—allow the difference in cost to accumulate with legal interest, less the expense of one coat of paint a year, and by the time the stone is ruined the iron will not only have cleared itself and stand on the balance sheet at a profit, but be in prime condition for continued service. On any much-travelled street a marble front soon becomes rusty and discolored with dust and rain. An iron front kept properly painted appears periodically in a new dress, and is always clean and bright. Other things being equal, place two merchants respectively in a stone front and an iron front store, side by side, and he in the clean, bright, attractive front will do the most business, and can afford to pay the largest rent. A stone front soon becomes discolored and dirty, and shows almost as many different soiled colors as there are different pieces of stone, caused by the chemical ingredients in the stone striking to the surface. An iron front reveals no joints, and looks as though it were cut out of one solid block and of one even color. Every time it is painted it looks new. More than one white marble front now regularly receives a coat of white paint to keep it white, because without the paint they looked dark and dingy alongside of their neighboring white iron fronts.

A great deal has been written about the color to paint iron

work. Iron being a material which requires a coating of lead
and oil, it is proper to give it any color that good taste may
suggest. The color will often be regulated by the color and
hue of adjoining buildings or other surroundings. Because
marble is white or sandstone brown, the painting of iron work in
these colors must not be prohibited. What is to be condemned
is the graining of iron in imitation of marble, and sanding in
imitation of stone. Tints and colors and gilding produce rich
and sparkling effects, but great care and exceeding good taste
must be exercised or failure will be the result. The best pig-
ments must be used, or the colors, exposed to the air and sun,
will fade rapidly—and the best do fade—and leave the front
shabby. Wherever practicable, iron work should be painted
inside as well as out, without delay. Particular care in this
respect should be given to all parts put together in pieces, as
cornices, trusses, etc. These should have their joints well
painted before being bolted or riveted together. Painting on
the inside, however, applies only to the shell parts. Columns
cannot be painted on the inside, nor do they need it. Column
stands over column with an intervening plate; the very construction makes of the inside of a column an almost air-tight
chamber, where the air is always dry. No oxidation takes
place under these conditions, and so no paint is necessary.
The inside of a column is covered with a coating of foundry
sand, which clings to it for ages. On the shell work, when
the paint has fairly reached every crevice, these parts too
become air-tight, and paint only becomes requisite on the
outside, and to brighten up the color. In applying ornaments,
such as leaves of capitals, etc., not only should the ornaments
themselves be first thoroughly painted, but the screws which
fasten the ornament to the main work should be dipped in
paint as well. After drilling a hole in iron, the burs around
the hole should always be filed away, so that no streaks of rust
from rain-water down the face of the building will tell of

carelessness in this respect. A lack of care in such little
matters often causes the greatest annoyance, and has been the
chief reason why iron fronts have had to be painted more
often during the ensuing few years of their erection than
afterward. Some fronts in a dark color have only been
painted for intervals of five years during the past twenty
years, and previous to that did not average more than once in
two years. For the first coating of iron nothing is superior to
oxide of iron mixed with oil, or what is known as metallic
paint.

On the manufacturer depends the artistic appearance of an
iron building, as well as its durability. The material is capable of receiving the sharpest kind of lines. But to secure
under-cuttings and that certain crispness necessary to the
proper effect, particularly of carved work, requires a combined
technical knowledge of architectural detail, of artistic pattern-
work, and of foundry moulding, and, withal, a business pride
and reputation. An architect may design a front, but its execution is beyond his control, and its effect, whether very ornate
or very plain, may be entirely spoiled by falling into the hands
of incompetent mechanics. Between the fronts of to-day and
those erected not many years ago there is a perceptible improvement. The artistic working-up of the material is better
understood. After years of alterations and comparison, boldness and good proportion in every part has been obtained.
The greatest possible caution should be exercised in awarding
contracts, and the difference of any moderate sum should never
permit the giving of work to parties who are lacking in experience or in knowledge or in facilities, or who habitually do
their work in a slovenly manner, or who are notoriously slow.
It is not always to the interest of an owner to give his work to
the lowest bidder. The grade of men in the iron business in
no wise differs from any other manufacturers, in that there are
some whose productions are superior and intrinsically worth

more than the like made by others. The thousand items of intricate detail about a job of iron work, which go to make up a complete whole, each of which requires the direct supervision of competent principals, but faintly tells of the constant and unwearying watchfulness that must be given to ensure good results.

Much has been said against iron from misconception. It is exceedingly difficult in the minds of most writers and talkers, who use sweeping denunciations and citations against iron, to separate wrought iron and cast iron in their respective endurance against weather. Wrought iron rapidly oxidizes when exposed to the atmosphere, and goes to decay. Cast iron, on the contrary, slowly oxidizes in damp situations ; rust does not scale from it, and the oxidation, when formed, is of a much less dangerous kind than on wrought iron. A coating of paint will counteract whatever tendency cast iron has to rust when exposed.

Whatever has been done in iron which deserves censure from critics can be remedied. Let it not be forgotten that the *material* is not at fault, but the *workmanship*. Iron can be made to imitate anything perfectly. Men who have said most against iron have been they who knew the least about it. Arguments have been made that iron is a sham, but a stone building is a greater sham, because it leads one to believe that it is all stone, when, in fact, it is nothing but a veneer set up against a brick wall.

The adaptability of all building materials depends principally upon their property of resisting the destroying influences of the atmospheric air, be these influences either mechanical or chemical. The objection to brown stone for buildings is that it is porous, and rains penetrate it ; the water freezes, and in expanding scales off the exterior layer, and a rapid decay is the result. Marble is denser, but every rain-storm dissolves a thin film of its surface. A bowl of water collected from the rain that has touched a marble front will be found by chemical test to be so charged with carbonate of lime as to be unfit for purposes for which rain-water is required. The effect is that the sharp edges of the architectural details become blunted, and gradually wear away. In marble there is carbonate of iron, which absorbs oxygen from the air, and then presents itself in yellowish spots, which gradually turn brown or black. Granite, which is the best building stone in the world, when subjected to strong heat cracks and splits off in flakes, and crumbles like dry plaster.

When iron fronts were first introduced, it was strenuously asserted by some that expansion and contraction would dislocate the joints and render a building unsafe. An examination of any of the numerous cast iron structures which, for a number of years, have been exposed to every change of atmospheric temperature without, and to the heat of steam-boilers, etc., within, will show everything unchanged. This proves that the temperature of our climate throughout its utmost range, from the greatest heat to the greatest cold, exerts upon it no appreciable effect. Events have also proven in the cases of burning of storehouses filled with combustible goods that cast iron fronts are absolutely fire-proof, and will neither warp nor crack nor fall down, unless the entire building falls, pulling the front with it. Only let it be remembered, that, in addition to a high and intense heat, the use of a blast is required to reduce cast iron to a molten state, and the ability of iron fronts to stand heat will be readily understood. They are also perfectly safe during thunder-storms; the metal presents so great a mass to the over-charged clouds, so as to become a huge conductor in itself, and silently conveys all the electricity to the earth. In them the intensity current is instantly diffused throughout the entire mass, and changed into a current of quantity, thus obviating all danger from disruptive discharges. Iron fronts have stood erect in cases where the side brick walls were entirely thrown down and demolished by the elements.

A front of iron is usually laid down and fitted together complete in the manufactory previous to erection at the building. It can be transported to any distance to the place of erection, and put together with wonderful rapidity, and at all seasons of the year. It takes up less space than any other material, and so enlarges the interior of a building. When it becomes desirable to tear down the building itself, to make way for other improvements, the iron front may be taken to pieces, without injury to any of its parts, and be re-erected elsewhere with the same perfection as at first. Instead of destruction, there need be a removal only.

Iron has in its favor unequalled advantages of ornament, strength, lightness of structure, facility of erection, durability, economy, incombustibility, and ready renovation. In iron, as in other materials, must ever be observed those undeviating laws of proportion, and rules deduced from a refined analysis of what is suitable in the highest degree to the end proposed. There is not a structure erected anywhere but adds its quantum to the good or bad impressions to be directly stamped upon the public mind. Thus every one who builds is unwittingly enhancing or deteriorating the taste of the masses, and the aggregate result of this is a thing not to be overestimated. It behooves the general use and careful treatment of a material which allows greater architectural effect, in proportion to the outlay of money, than any other. The uses and requirements and values of buildings are changing every day, and iron in its architectural application is to fulfil future requirements such as in the past it has but limitedly supplied. In our new and growing country, the dollars saved on one building are required for the erection of another. It is primarily a duty for every builder to do the most with his money, and the most for Art. When the public become thoroughly acquainted with the advantages iron possesses as a building material, it is confidently predicted that for superior buildings of all kinds it will receive a general preference to granite, marble, sandstone, or brick.

ASHLER FRONTS.

Thin plates of cast iron used as a facing to brick walls give a very good effect. The plates need not be over three-sixteenths of an inch thick, and each should have two eyes cast on the back to anchor in the walls. The plates should not be bolted together, and a proper allowance must be made in the joints for the shrinkage and settling of the brick wall; this can be done by using strips of rubber at each joint, or strips of wood, which are afterwards removed. At the joints the plates should have lips turning upwards and downwards, so that rain cannot beat in, or sweat or moisture from the inside trickle out.

The outer surface of the plates should be roughened with very small uniform corrugations so as to be more pleasing to the eye, and to avoid the shining or glistening effect produced by oil paint on a smooth surface. This roughening also conceals the slight irregularities and warpings so painfully apparent when the light strikes at certain angles on the surfaces of plain castings. The glare of paint is thus deadened, and gives a chiaroscuro effect (light and shade), very closely resembling stone, and more beautiful than the tool marks on stone surfaces. These rounded ridges, and correspondingly rounded hollows between, should be sufficiently fine to prevent their being noticeable at ordinary distances, and yet sufficiently coarse to avoid being filled with paint—say eight ridges and hollows to an inch; a ridge and hollow together making about one-eighth of an inch in width. The corrugations may be done either on the wood patterns, or by means of a suitable moulder's tool or "slicker," grooved and ridged to exactly the extent desired, and which is operated successfully and rapidly in producing the desired reverse grooves and ridges on the surface of the mould.

ROLLED IRON BEAMS AND CHANNELS.

[NEW JERSEY STEEL AND IRON CO.]

The following tables give the principal data relating to rolled beams which are required in practice.

The safe loads given in the tables are those which can be carried in addition to the weight of the beam itself when the beams are supported at both ends and the load uniformly distributed over the length of the beam, and are such as would bring a maximum strain upon the iron of 12,000 lbs. per square inch, this being about one-quarter the breaking weight of wrought iron.

As, in building, the admissible deflection of beams is limited by the amount which would cause the plastering of ceilings to crack, the tables have a cross-line dividing them at the length of span at which this is found to occur; the lengths and loads above the line being proper for plastered ceilings, and those below to be used only when this consideration does not enter. The limit of deflection thus allowed is one-thirtieth of an inch to the foot of span.

FIRE-PROOF FLOORS.

The load upon the beams of fire-proof floors, with four-inch brick arches levelled up with concrete between the beams, in buildings used for offices, assemblages of people, or storage of light goods, may, ordinarily, be taken at 70 lbs. per square foot of floor for the weight of the arches, concrete, ceiling, and flooring; and at 80 lbs. per square foot additional for a variable load equal to the weight of a crowd of people; making a total load of 150 lbs. per square foot of floor, in addition to the weight of the beams.

For street bridges for general public traffic, a load in addition to the weight of the structure, of 80 lbs. per square foot, may be taken.

For the floors of dwellings............	40 lbs.
Churches, theatres, and ball-rooms.....	80 "
Hay-lofts..........................	80 "
Storage of grain....................	100 "
Warehouses and general merchandise...	250 "
Factories.....................200 to 400 "
Snow 30 inches deep................	16 "
Maximum pressure of wind..........	50 "
Brick walls, per cubic foot..........	112 "
Masonry walls,   "   .....116 to 144 "

USE OF THE TABLES.—What beams would be required for a floor 50 ft. by 21 ft. in the clear, to be used for offices, and therefore loaded to the extent of 150 lbs. per square foot, and what will be the total weight of iron?

Supposing that it is desired to make the brick arches about 4 ft. in span between the beams, we find this distance opposite 21 ft. span and under 150 lbs. per square foot, in the table for $10\frac{1}{2}''$ light beams. As this comes above the cross-line of the table, these beams could be used without injury to the plastering on account of deflection. The distance between the centres of beams being 4.1 ft., there would be required 13 beams. Allowing 8" bearing at each end of the beams, the total length of each would be 22' 4", the weight of which is 781.7 lbs., or, for the 13 beams, 10,162 lbs.

If a deeper beam is preferred, $12\frac{1}{4}''$ light, for instance, may be substituted, and, referring to the table for this beam, we find that for the above load and span they should be spaced 5.4 ft. apart, and there will, therefore, be but 10 beams required, the weight of which would be 9,300 lbs.

4-INCH LIGHT BEAM—30 LBS. PER YARD.

TRENTON BEAMS.				Length, in inches	Weight, in lbs.	Length, in feet	Weight, in lbs.
				1	0.8	1	10
				2	1.7	2	20
Depth, 4 inches.				3	2.5	3	30
				4	3.3	4	40
Width of flanges, 2¼ inches.				5	4.2	5	50
				6	5.0	6	60
Thickness of stem, ¼ inch.				7	5.8		
				8	6.7		
Area of cross-section, 2.96 sq. inches.				9	7.5		
				10	8.3		
				11	9.2		

Distance between supports, in feet.	Safe uniformly distributed load, in tons of 2,000 lbs.	Deflection, in inches, under this load.	Weight in lbs.	Proper Distance, in feet, between Centres of Beams, per load in lbs., per square foot of—						
				100 lbs.	125 lbs.	150 lbs.	175 lbs.	200 lbs.	250 lbs.	300 lbs.
5	2.99	0.09	50.	12.0	9.6	8.0	6.8	6.0	4.8	4.0
6	2.48	0.13	60.	8.3	6.6	5 5	4.7	4.1	3.3	2.8
7	2.11	0.17	70.	6.0	4.8	4.0	3.4	3.0	2.4	2.0
8	1.84	0.23	80.	4.6	3.7	3.1	2.6	2.3	1.8	1.5
9	1.63	0.29	90.	3.6	2.9	2.4	2.1	1.8	1.4	1.2
10	1.45	0.36	100.	2.9	2.3	1.9	1.7	1.4	1.2	1.0
11	1.31	0.43	110.	2.4	1.9	1.6	1.4	1.2	1.0	0.8
12	1.19	0.51	120.	2.0	1.6	1.3	1.1	1.0	0.8	0.7
13	1 09	0.60	130.	1.7	1.3	1.1	1.0	0.8	0.7	0.6
14	1.00	0.70	140.	1.4	1.1	0.9	0.8	0.7	0.6	0 5
15	0.93	0.81	150.	1.2	1.0	0.8	0.7	0.6	0.5	0.4
16	0.86	0.90	160.	1.1	0.9	0.7	0.6	0.5	0.4	0.4
17	0.80	1.03	170.	0.9	0.8	0 6	0.5	0.4	0.4	0.3
18	0.75	1.16	180.	0.8	0.7	0.5	0.5	0.4	0.3	0.3
19	0.70	1.29	190.	0.7	0.6	0.5	0.4	0.3	0.3	0 2
20	0.65	1.43	200.	0.7	0.5	0.4	0 4	0.3	0.2	0.2
21	0.61	1.58	210.	0.6	0.5	0.4	0.3			
22	0.57	1.73	220.	0.5	0.4	0.3	0.3			
23	0.54	1.89	230.	0.5	0.4	0.3	0.2			
24	0.51	2.06	240.	0.4	0.3	0.3	0.2			
25	0.48	2.23	250.	0.4	0.3	0.3	0.2			
26	0.45	2.41	260.	0.3						
27	0.42	2.60	270.	0.3						
28	0.40	2.79	280.	0.3						
29	0.37	3.00	290.	0.3						
30	0.35	3.21	300.	0.2						

4-INCH HEAVY BEAM—37 LBS. PER YARD.

TRENTON BEAMS.		Length, in inches.	Weight in lbs.	Length, in feet.	Weight, in lbs.
	Depth, 4 inches.	1	1.0		
		2	2.0	1	12.3
		3	3.1	2	24.7
	Width of flanges, 3 inches.	4	4.1	3	37.0
		5	5.1	4	49.3
	Thickness of stem, $\frac{5}{16}$ inch.	6	6.1	5	61.7
		7	7.2	6	74.0
	Area of cross-section, 3.66 sq. inches.	8	8.2		
		9	9.3		
		10	10.3		
		11	11.3		

Distance between supports, in feet.	Safe uniformly distributed load, in tons of 2,000 lbs.	Deflection, in inches, under this load.	Weight in lbs.	Proper Distance, in feet, between Centres of Beams, for load in lbs., per square foot of—						
				100 lbs.	125 lbs.	150 lbs.	175 lbs.	200 lbs.	250 lbs.	300 lbs.
5	3 65	0.09	61.7	14.6	11.7	9.7	8.3	7.3	5.0	4 9
6	3.02	0.13	74.0	10.1	8.1	6.7	5.8	5 0	4.0	3.3
7	2.59	0.17	86.3	7 4	5.9	4.9	4.2	3.7	3.0	2.5
8	2.25	0.23	98.7	5.6	4.5	3.7	3.2	2.8	2.8	1.9
9	1.99	0.29	111.0	4.4	3.5	2.9	2.5	2.2	1.8	1.5
10	1.78	0.36	123.3	3.6	2.8	2.4	2.1	1.8	1.4	1.2
11	1.60	0.43	135.7	2.9	2.3	1.9	1.7	1.4	1.2	1.0
12	1.46	0.51	148.0	2.4	1 9	1.6	1.4	1.2	1 0	0.8
13	1.34	0 60	160.3	2.0	1.6	1.3	1.2	1.0	0.8	0.7
14	1.23	0.70	172.7	1.7	1.3	1.1	1.0	0.8	0.7	0.6
15	1.13	0.81	185.0	1.5	1 2	1.0	0.9	0.7	0.6	0.5
16	1.05	0.91	197.3	1.3	1.0	0.9	0.7	0.6	0.5	0.4
17	0.98	1.03	209 7	1.1	0.9	0.8	0.6	0.5	0.5	0.4
18	0.91	1.16	222.0	1.0	0.8	0.7	0.6	0.5	0.4	0.3
19	0.85	1.29	234.3	0.9	0.7	0.6	0.5	0.4	0.4	0.3
20	0.80	1.43	246.7	0.8	0.6	0.5	0.5	0.4	0.3	0.3
21	0.75	1.58	259.0	0.7	0.6	0.5	0.4	0.3		
22	0.70	1.73	271.3	0.6	0.5	0.4	0.4	0 3		
23	0.66	1.89	283.7	0.6	0.5	0.4	0.3	0.3		
24	0.62	2.06	296.0	0.5	0.4	0.3	0.3			
25	0.58	2.23	308.3	0.5	0.4	0.3	0.3			
26	0 55	2.41	320.7	0.4	0.3					
27	0.52	2.60	333.0	0.4	0.3					
28	0.48	2.79	345.3	0.3						
29	0.45	3.00	357.7	0.3						
30	0.43	3.21	370.0	0.3						

5-INCH LIGHT BEAM—30 LBS. PER YARD.

TRENTON BEAM.	Length, in inches.	Weight, in lbs.	Length, in feet.	Weight, in lbs.
Depth, 5 inches.	1	0.8	1	10
	2	1.7	2	20
	3	2.5	3	30
Width of flanges, 2¼ inches.	4	3.3	4	40
	5	4.2	5	50
	6	5.0	6	60
Thickness of stem, ¼ inch.	7	5.8		
	8	6.7		
	9	7.5		
Area of cross-section, 2.96 sq. inches.	10	8.3		
	11	9.2		

Distance between supports, in feet.	Safe uniformly distributed load, in tons of 2,000 lbs.	Deflection, in inches, under this load.	Weight, in lbs.	Proper Distance, in feet, between Centres of Beams, for load, in lbs., per square foot of—						
				100 lbs.	125 lbs.	150 lbs.	175 lbs.	200 lbs.	250 lbs.	300 lbs.
5	3.85	0.07	50.	15.4	12.3	10.3	8.8	7.7	6.2	5.1
6	3.19	0.10	60.	10.6	8.5	7.1	6.1	5.3	4.2	3.5
7	2.73	0.14	70.	7.8	6.3	5.2	4.5	3.9	3.1	2 6
8	2.38	0.18	80.	5.9	4.8	4.0	3.4	3.0	2 4	2.0
9	2.10	0.23	90.	4.7	3.7	3.1	2.7	2.3	1.9	1.6
10	1.88	0.28	100.	3.8	3.0	2.5	2.2	1.9	1.5	1.3
11	1.70	0.34	110.	3 1	2.5	2.1	1.8	1.5	1.2	1.0
12	1.55	0.41	120.	2.6	2.1	1.7	1.5	1.3	1.0	0.9
13	1.42	0.48	130.	2.2	1.8	1.5	1.3	1.1	0.9	0.7
14	1.31	0.56	140.	1.9	1.5	1.3	1.1	0.9	0.8	0.6
15	1.21	0.64	150.	1.6	1.3	1.1	1 0	0.8	0.6	0.5
16	1.13	0.73	160.	1.4	1.1	1.0	0.8	0.7	0.6	0.5
17	1.05	0.82	170.	1.2	1.0	0.8	0.7	0.6	0.5	0.4
18	0.98	0.92	180.	1.1	0.9	0.7	0.6	0.5	0.4	0.4
19	0.92	1.03	190.	1 0	0.8	0.7	0.6	0.5	0.4	0.3
20	0.87	1.14	200.	.9	0 7	0.5	0.5	0.4	0.3	0.3
21	0.82	1.26	210.	0.8	0.7	0.5	0.5	0.4	0.3	0.3
22	0.77	1.38	220.	0.7	0.6	0.5	0.4	0.3	0.3	
23	0.73	1.51	230.	0.6	0 5	0.4	0.4	0 3		
24	0.69	1.65	240.	0.6	0.5	0.4	0.4	0.3		
25	0.65	1.79	250.	0.5	0.4	0.3	0.3			
26	0.61	1.93	260.	0.4	0.3	0.3				
27	0.58	2.08	270.	0.4	0.3	0.3				
28	0.55	2.24	280.	0.4	0.3	0.3				
29	0.52	2.40	290.	0.3						
30	0.50	2.57	300.	0.3						

5-INCH HEAVY BEAM—40 LBS. PER YARD.

TRENTON BEAM.		Length, in inches.	Weight, in lbs.	Length, in feet.	Weight, in lbs.
	Depth, 5 inches.	1	1.1	1	13.3
		2	2.2	2	26.7
		3	3.3	3	40.0
	Width of flanges, 3 inches.	4	4.4	4	53.3
		5	5.5	5	66.7
		6	6.7	6	80.0
	Thickness of stem, 5-16 inch.	7	7.8		
		8	8.9		
		9	10.0		
	Area of cross-section, 3.90 sq. inches.	10	11.1		
		11	12.2		

Distance between supports, in feet.	Safe uniformly distributed load, in tons of 2,000 lbs.	Deflection, in inches, under this load.	Weight, in lbs.	Proper Distance, in feet, between Centres of Beams, for load, in lbs., per square foot of—						
				100 lbs.	125 lbs.	150 lbs.	175 lbs.	200 lbs.	250 lbs.	300 lbs.
5	4.87	0.07	66.7	19.5	15.6	13.0	11.1	9.8	7.8	6.5
6	4.05	0.10	80.0	13.5	10.8	9.0	7.7	6.8	5.4	4.5
7	3.46	0.14	93.3	9.9	7.9	6.6	5.7	4.9	4.0	3.3
8	3.02	0.18	106.7	7.5	6.0	5.0	4.3	3.8	3.0	2.5
9	2.67	0.23	120.0	5.9	4.7	4.0	3.4	3.0	2.4	2.0
10	2.39	0.28	133.3	4.8	3.8	3.2	2.7	2.4	1.9	1.6
11	2.16	0.34	146.7	3.9	3.1	2.6	2.2	2.0	1.6	1.3
12	1.97	0.41	160.0	3.3	2.6	2.2	1.9	1.6	1.3	1.1
13	1.80	0.48	173.3	2.8	2.3	1.8	1.6	1.4	1.1	0.9
14	1.66	0.56	186.7	2.4	1.9	1.6	1.3	1.2	1.0	0.8
15	1.54	0.64	200.0	2.0	1.6	1.4	1.1	1.0	0.8	0 7
16	1.43	0.73	213.3	1.8	1.4	1.2	1.0	0.9	0.7	0.6
17	1.33	0.82	226.7	1.6	1.3	1.1	0.9	0.8	0.6	0.5
18	1.24	0.92	240.0	1.4	1.1	1.0	0.8	0.7	0.6	0.5
19	1.16	1.03	253.3	1.2	1.0	0.8	0.7	0.6	0.5	0.4
20	1.09	1.14	266.7	1.1	0.9	0.7	0.6	0.5	0.4	0.4
21	1.03	1.26	280.0	1.0	0.8	0.7	0.6	0.5	0.4	0.3
22	0.97	1.38	293.3	0.9	0.7	0.6	0.5	0.4	0.4	0.3
23	0.91	1.51	306.7	0.8	0.6	0.5	0.5	0.4	0.3	0.3
24	0.86	1.65	320.0	0.7	0.6	0.5	0.4	0.3	0.3	
25	0.82	1.79	333.3	0.6	0.5	0.4	0.3	0.3		
26	0.77	1.93	346.7	0.6	0.5	0.4	0.3	0.3		
27	0.73	2.08	360.0	0.5	0.4	0.3	0.3			
28	0.69	2.24	373.3	0.5	0.4	0.3	0.3			
29	0.65	2.40	386.7	0.4	0.3	0.3				
30	0.62	2.57	400.0	0.4	0.3	0.3				

6-INCH LIGHT BEAM—40 LBS. PER YARD.

Trenton Beam.		Length, in inches.	Weight, in lbs.	Length, in feet.	Weight, in lbs.
	Depth, 6 inches.	1	1.1	1	13.3
		2	2.2	2	26.7
		3	3.3	3	40.0
	Width of flanges, 3 inches.	4	4.4	4	53.3
		5	5.5	5	66.7
		6	6.7	6	80.0
	Thickness of stem, ¼ inch.	7	7.8		
		8	8.9		
		9	10.0		
	Area of cross-section, 4.01 sq. inches.	10	11.1		
		11	12.2		

Distance between supports, in feet.	Safe uniformly distributed load, in tons of 2,000 lbs.	Deflection, in inches, under this load.	Weight in lbs.	Proper Distance, in feet, between Centres of Beams, for load, in lbs., per square foot of—						
				100 lbs.	125 lbs.	150 lbs.	175 lbs.	200 lbs.	250 lbs.	300 lbs.
5	5.18	0.06	66.7	20.7	16.6	13.8	11.8	10.4	8.3	6.9
6	5.18	0 08	80.0	17.3	13.8	11.5	9.9	8.6	6.9	5.8
7	4.42	0.12	93.3	12.6	10.1	8.4	7 2	6.3	5.1	4.2
8	3.86	0.15	106.7	9.6	7.7	6.4	5.4	4.8	3.9	3.2
9	3.42	0.19	120.0	7.6	6.0	5.1	4.3	3 8	3.0	2.5
10	3.06	0.24	133.3	6.1	4.9	4.1	3.5	3.0	2.4	2.0
11	2.77	0.29	146.7	5.0	4.0	3.4	2.9	2.5	2.0	1.7
12	2.53	0.34	160.0	4.2	3.4	2.8	2.4	2.1	1.7	1.4
13	2.32	0 40	173.3	3.6	2.9	2.4	2.1	1.8	1.4	1.2
14	2.14	0.47	186.7	3.1	2.5	2.0	1.8	1.5	1.2	1.0
15	1.99	0 54	200.0	2.6	2.1	1.8	1.5	1.3	1.0	0.9
16	1.85	0.61	213.3	2 3	1.8	1.5	1.3	1.1	0.9	0.8
17	1.73	0.69	226.7	2.0	1.6	1.4	1.1	1.0	9.8	.7
18	1.62	0.77	240.0	1.8	1.4	1.2	1.0	0.9	0.8	0.6
19	1.52	0.86	253.3	1.6	1.3	1.1	0.9	0.8	0.6	0.5
20	1.43	0.95	266.7	1.4	1.1	1.0	0.8	0.7	0.6	0.5
21	1.35	1.05	280.0	1.3	1.0	0.9	0.7	0.6	0.5	0.4
22	1.28	1.15	293.3	1.2	1.0	0.8	0.7	0.6	0.5	0.4
23	1.21	1.26	306.7	1.1	0.9	0.8	0.6	0.5	0.4	0.4
24	1.14	1.37	320.0	1.0	0.8	0.7	0.6	0.5	0.4	0.3
25	1.08	1.49	333.3	0.9	0.7	0.6	0.5	0.4	0.4	0.3
26	1.03	1.61	346.7	0.8	0.7	0.6	0.5	0.4	0.3	0.3
27	0.98	1.74	360.0	0.7	0.6	0.5	0.4	0.3	0.3	
28	0.93	1.87	373.3	5.7	0.6	0.5	0.4	0.3	0.3	
29	0.89	2.00	386.7	0.6	0.5	0.4	0.3	0.3		
30	0.84	2.14	400.0	0.6	0.5	0.4	0.3	0.3		

7

6-INCH HEAVY BEAM—50 LBS. PER YARD.

Trenton Beam.		Length, in inches.	Weight, in lbs.	Length, in feet.	Weight, in lbs.
	Depth, 6 inches.	1	1.4	1	16.7
		2	2.7	2	33.3
		3	4.1	3	50.0
	Width of flanges, 3¼ inches.	4	5.4	4	66.7
		5	6.8	5	83.3
		6	8.3	6	100.0
	Thickness of stem, 0.3 inch.	7	9.5		
		8	10.9		
		9	12.3		
	Area of cross-section, 4.91 sq. inches.	10	13.7		
		11	15.1		

Distance, between supports, in feet.	Safe uniformly distributed load, in tons of 2,000 lbs.	Deflection, in inches, under this load.	Weight, in lbs.	Proper Distance, in feet, between Centres of Beams, for load, in lbs., per square foot of—						
				100 lbs.	125 lbs.	150 lbs.	175 lbs.	200 lbs.	250 lbs.	300 lbs.
5	6.36	0.06	83.3	25.4	20.4	17.0	14.5	12.7	10.2	8.5
6	6.35	0.08	100.0	21.2	16.9	14.1	12.1	10.6	8.5	7.1
7	5.43	0.12	116.7	15.5	12.4	10.3	8.9	7.8	6.2	5.2
8	4.73	0.15	133.3	11.8	9.4	7.9	6.8	5.9	4.7	3.9
9	4.19	0.19	150.0	9.3	7.4	6.2	5.3	4.6	3.7	3.1
10	3.76	0.24	166.7	7.5	6.0	5.0	4.3	3.7	3.0	2.5
11	3.40	0.29	183.3	6.2	4.9	4.1	3.5	3.1	2.5	2.1
12	3.10	0.34	200.0	5.2	4.2	3.4	3.0	2.6	2.1	1.7
13	2.85	0.40	216.7	4.4	3.5	2.9	2.5	2.2	1.8	1.5
14	2.63	0.47	233.3	3.7	3.0	2.5	2.1	1.8	1.5	1.2
15	2.43	0.54	250.0	3.2	2.6	2.2	1.8	1.6	1.3	1.1
16	2.27	0.61	266.7	2.8	2.2	1.9	1.6	1.4	1.1	0.9
17	2.12	0.69	283.3	2.5	2.0	1.7	1.4	1.2	1.0	0.8
18	1.98	0.77	300.0	2.2	1.7	1.5	1.3	1.1	0.9	0.7
19	1.86	0.86	316.7	1.9	1.5	1.3	1.1	0.9	0.8	0.6
20	1.75	0.95	333.3	1.7	1.4	1.2	1.0	0.8	0.7	0.6
21	1.65	1.05	350.0	1.6	1.3	1.1	0.9	0.8	0.6	0.5
22	1.56	1.15	366.7	1.4	1.1	0.9	0.8	0.7	0.6	0.5
23	1.48	1.26	383.3	1.3	1.0	0.9	0.7	0.6	0.5	0.4
24	1.40	1.37	400.0	1.2	1.0	0.8	0.7	0.6	0.5	0.4
25	1.33	1.49	416.7	1.1	0.9	0.7	0.6	0.5	0.4	0.4
26	1.26	1.61	433.3	1.0	0.8	0.7	0.6	0.5	0.4	0.3
27	1.20	1.74	450.0	0.9	0.7	0.6	0.5	0.4	0.4	0.3
28	1.14	1.87	466.7	0.8	0.6	0.5	0.4	0.4	0.3	0.3
29	1.08	2.00	483.3	0.7	0.6	0.5	0.4	0.3	0.3	0.2
30	1.03	2.14	500.0	0.6	0.5	0.4	0.3	0.3	0.2	0.2

7-INCH BEAM—60 LBS. PER YARD.

TRENTON BEAM.		Length, in inches.	Weight, in lbs.	Length, in feet.	Weight, in lbs.
	Depth, 7 inches.	1	1.7	1	20
		2	3.3	2	40
		3	5.0	3	60
	Width of flanges, 3¼ inches.	4	6.7	4	80
		5	8.3	5	100
		6	10.0	6	120
	Thickness of stem, ⅝ inch.	7	11.7		
		8	13.3		
		9	15.0		
	Area of cross-section,	10	16.7		
	5.84 sq. inches.	11	18.3		

Distance between supports, in feet.	Safe uniformly distributed load, in tons of 2,000 lbs.	Deflection, in inches, under this load.	Weight, in lbs.	Proper Distance, in feet, between Centres of Beams, for load, in lbs., per square foot of—						
				100 lbs.	125 lbs.	150 lbs.	175 lbs.	200 lbs.	250 lbs.	300 lbs.
5	8.45	0.05	100	33.8	27.0	22.5	19.3	16.9	13.5	11.2
6	8.44	0.07	120	28.1	22.5	18.7	16.1	14.0	11.2	9.4
7	7.21	0.10	140	20.6	16.5	13.7	11.8	10.3	8.2	6.9
8	6.29	0.13	160	15.7	12.6	10.5	9.0	7.9	6.3	5.2
9	5.57	0.17	180	12.4	9.9	8.3	7.1	6.2	5.0	4.1
10	5.00	0.20	200	10.0	8.0	6.7	5.7	5.0	4.0	3.3
11	4.53	0.25	220	8.2	6.6	5.5	4.7	4.1	3.3	2.7
12	4.13	0.29	240	6.9	5.5	4.6	4.0	3.4	2.8	2.3
13	3.79	0.35	260	5.8	4.6	3.9	3.3	2.9	2.3	1.9
14	3.50	0.40	280	5.0	4.0	3.3	2.9	2.5	2.0	1.7
15	3.25	0.46	300	4.3	3.4	2.9	2.5	2.1	1.7	1.4
16	3.03	0.52	320	3.8	3.0	2.5	2.2	1.9	1.5	1.3
17	2.83	0.59	340	3.3	2.7	2.2	1.9	1.6	1.3	1.1
18	2.65	0.66	360	2.9	2.3	2.0	1.7	1.4	1.2	1.0
19	2.49	0.74	380	2.6	2.0	1.7	1.5	1.3	1.0	0.9
20	2.35	0.82	400	2.3	1.8	1.6	1.3	1.1	0.9	0.8
21	2.22	0.90	420	2.1	1.7	1.4	1.2	1.0	0.8	0.7
22	2.10	0.99	440	1.9	1.5	1.3	1.1	0.9	0.8	0.6
23	1.99	1.08	460	1.7	1.4	1.1	1.0	0.8	0.7	0.6
24	1.88	1.17	480	1.6	1.3	1.0	0.9	0.8	0.6	0.5
25	1.79	1.27	500	1.4	1.1	0.9	0.8	0.7	0.6	0.5
26	1.70	1.38	520	1.3	1.0	0.9	0.7	0.6	0.5	0.4
27	1.62	1.49	540	1.2	1.0	0.8	0.7	0.6	0.5	0.4
28	1.54	1.60	560	1.1	0.9	0.7	0.6	0.5	0.4	0.3
29	1.47	1.72	580	1.1	0.9	0.7	0.6	0.5	0.4	0.3
30	1.40	1.84	600	1.0	0.8	0.7	0.6	0.5	0.4	0.3

8-INCH LIGHT BEAM—65 LBS. PER YARD.

TRENTON BEAM.		Length, in inches.	Weight, in lbs.	Length, in feet.	Weight, in lbs.
	Depth, 8 inches.	1	1.81	1	21.7
		2	3.61	2	43.3
		3	5.42	3	65.0
	Width of flanges, 4 inches.	4	7.23	4	86.7
		5	9.03	5	108.3
		6	10.83	6	130.0
	Thickness of stem, .3 inch.	7	12.64		
		8	14.45		
		9	16.25		
	Area of cross-section, 6.37 sq. inches.	10	18.06		
		11	19.86		

Distance between supports, in feet.	Safe uniformly distributed load, in tons of 2,000 lbs.	Deflection, in inches, under this load.	Weight, in lbs.	Proper Distance, in feet, between Centres of Beams, for load, in lbs., per square foot of—						
				100 lbs.	125 lbs.	150 lbs.	175 lbs.	200 lbs.	250 lbs.	300 lbs.
6	11.18	0.06	130.0	37.3	29.8	24.9	21.3	18.6	14.9	12.4
7	9.57	0.08	151.7	27.3	21.8	18.2	15.6	13.6	10.9	9.1
8	8.35	0.11	173.3	20.9	16.7	13.9	11.9	10.4	8.3	7.0
9	7.40	0.14	195.0	16.4	13.2	10.9	9.4	8.2	6.6	5.5
10	6.64	0.18	216.7	13.3	10.6	8.9	7.6	6.6	5.3	4.4
11	6.02	0.22	238.3	10.9	8.7	7.3	6.2	5.4	4.3	3.6
12	5.49	0.26	260.0	9.1	7.3	6.1	5.2	4.5	3.6	3.0
13	5.05	0.30	281.7	7.8	6.2	5.2	4.5	3.9	3.1	2.6
14	4.67	0.35	303.3	6.7	5.3	4.5	3.8	3.3	2.7	2.2
15	4.34	0.40	325.0	5.8	4.6	3.9	3.3	2.9	2.3	1.9
16	4.04	0.46	346.7	5.0	4.0	3.3	2.9	2.5	2.0	1.7
17	3.79	0.52	368.3	4.4	3.5	2.9	2.5	2.2	1.8	1.5
18	3.55	0.58	390.0	3.9	3.1	2.6	2.2	1.9	1.6	1.3
19	3.35	0.64	411.7	3.5	2.8	2.3	2.0	1.7	1.4	1.2
20	3.16	0.71	433.3	3.2	2.6	2.1	1.8	1.6	1.3	1.1
21	2.99	0.79	455.0	2.8	2.3	1.9	1.6	1.4	1.1	1.0
22	2.83	0.86	476.7	2.6	2.1	1.7	1.5	1.3	1.0	0.9
23	2.68	0.94	498.3	2.3	1.9	1.5	1.3	1.1	0.9	0.8
24	2.55	1.03	520.0	2.1	1.7	1.4	1.2	1.0	0.8	0.7
25	2.43	1.12	541.7	1.9	1.5	1.3	1.1	0.9	0.8	0.6
26	2.31	1.21	563.3	1.8	1.4	1.2	1.0	0.9	0.7	0.6
27	2.20	1.30	585.0	1.7	1.3	1.1	1.0	0.8	0.7	0.6
28	2.10	1.40	606.7	1.5	1.2	1.0	0.9	0.8	0.6	0.5
29	2.01	1.50	628.3	1.4	1.1	0.9	0.8	0.7	0.6	0.5
30	1.93	1.61	650.0	1.3	1.0	0.9	0.7	0.6	0.5	0.4
31	1.84	1.72	671.7	1.2	1.0	0.8	0.7	0.6	0.5	0.4
32	1.76	1.83	693.3	1.1	0.9	0.7	0.6	0.5	0.4	0.3
33	1.69	1.94	715.0	1.0	0.8	0.7	0.6	0.5	0.4	0.3
34	1.62	2.06	736.7	0.9	0.7	0.6	0.5	0.5	0.4	0.3
35	1.55	2.18	758.0	0.8	0.7	0.5	0.5	0.4	0.3	0.2

8-INCH HEAVY BEAM—80 LBS. PER YARD.

TRENTON BEAM.	Length, in inches.	Weight, in lbs.	Length, in feet.	Weight, in lbs.
	1	2.2	1	26.7
Depth, 8 inches.	2	4.4	2	53.3
	3	6.7	3	80.0
Width of flanges, 4½ inches.	4	8.9	4	106.7
	5	11.1	5	133.3
	6	13.3	6	160.0
Thickness of stem, ⅝ inch.	7	15.6		
	8	17.8		
	9	20.0		
Area of cross-section, 8.03 sq. inch.	10	22.2		
	11	24.4		

Distance between supports, in feet	Safe uniformly distributed load, in tons of 2,000 lbs.	Deflection, in inches, under this load.	Weight in lbs.	Proper Distance, in feet, between Centres of Beams, for load, in lbs., per square foot of—						
				100 lbs.	125 lbs.	150 lbs.	175 lbs.	200 lbs.	250 lbs.	300 lbs.
6	9.27	0.05	160.0	30.9	24.7	20.6	17.7	15.4	12.4	10.3
7	9.25	0.08	186.7	26.4	21.1	17.6	15.1	13.2	10.6	8.8
8	9.23	0.11	213.3	23.1	18.5	15.4	13.2	11.5	9.2	7.7
9	9.21	0.14	240.0	20.5	16.4	13.6	11.7	10.2	8.2	6.8
10	8.27	0.18	266.7	16.5	13.2	11.0	9.4	8.2	6.6	5.5
11	7.40	0.22	293.3	13.6	10.9	9.1	7.8	6.8	5.4	4.5
12	6.84	0.26	320.0	11.4	9.1	7.6	6.5	5.7	4.6	3.8
13	6.29	0.30	346.7	9.7	7.7	6.4	5.5	4.8	3.9	3.2
14	5.81	0.35	373.3	8.3	6.6	5.5	4.7	4.1	3.3	2.7
15	5.40	0.40	400.0	7.2	5.7	4.8	4.1	3.6	2.9	2.4
16	5.04	0.46	426.7	6.3	5.0	4.2	3.6	3.1	2.5	2.1
17	4.71	0.52	453.3	5.5	4.4	3.7	3.2	2.7	2.2	1.8
18	4.43	0.58	480.0	4.9	3.9	3.3	2.8	2.4	1.9	1.6
19	4.17	0.64	506.7	4.4	3.5	2.9	2.5	2.2	1.8	1.5
20	3.93	0.71	533.3	3.9	3.1	2.6	2.2	1.9	1.6	1.3
21	3.72	0.79	560.0	3.5	2.8	2.3	2.0	1.7	1.4	1.2
22	3.52	0.86	586.7	3.2	2.6	2.1	1.8	1.6	1.3	1.1
23	3.34	0.94	613.3	2.9	2.3	1.9	1.7	1.4	1.2	1.0
24	3.18	1.03	640.0	2.6	2.1	1.7	1.5	1.3	1.1	0.9
25	3.03	1.12	666.7	2.4	1.9	1.6	1.4	1.2	1.0	0.8
26	2.88	1.20	693.3	2.2	1.8	1.5	1.2	1.1	0.9	0.7
27	2.75	1.30	720.0	2.0	1.6	1.3	1.1	1.0	0.8	0.7
28	2.63	1.40	746.7	1.8	1.5	1.2	1.0	0.9	0.7	0.6
29	2.51	1.50	773.3	1.7	1.4	1.1	1.0	0.8	0.7	0.6
30	2.40	1.61	800.0	1.6	1.3	1.1	0.9	0.8	0.6	0.5
31	2.30	1.72	826.7	1.5	1.2	1.0	0.9	0.7	0.6	0.5
32	2.20	1.83	853.3	1.4	1.1	0.9	0.8	0.7	0.6	0.4
33	2.11	1.94	880.0	1.3	1.0	0.9	0.7	0.6	0.5	0.4
34	2.02	2.06	906.7	1.2	0.9	0.8	0.7	0.6	0.5	0.4
35	1.93	2.19	933.3	1.1	0.9	0.7	0.6	0.5	0.4	0.3

9-INCH LIGHT BEAM—70 LBS. PER YARD.

Trenton Bea·.	Length, in inches.	Weight, in lbs.	Length, in feet.	Weight, in lbs.
Depth, 9 inches.	1	1.9	1	23.3
	2	3.9	2	46.7
Width of flanges, 3½ inches.	3	5.8	3	70.0
	4	7.8	4	93.3
	5	9.7	5	116.7
Thickness of stem, .3 inch.	6	11.7	6	140.0
	7	13.6		
	8	15.6		
Area of cross-section, 6.53 sq. inches.	9	17.5		
	10	19.4		
	11	21.4		

Distance between supports, in feet.	Safe uniformly distributed load, in tons of 2,000 lbs.	Deflection, in inches, under this load.	Weight in lbs.	Proper Distance, in feet, between Centres of Beams, for load. in lbs., per square foot of—						
				100 lbs.	125 lbs.	150 lbs.	175 lbs.	200 lbs.	250 lbs.	300 lbs.
6	9.43	0.05	140.0	31.4	25.1	21.0	17.9	15.7	12.6	10.5
7	9.42	0.07	163.3	26.9	21.5	17.9	15.3	13.4	10.8	9.0
8	9.41	0.10	186.7	23.5	18.8	15.7	13.4	11.7	9.4	7.8
9	8.34	0.13	210.0	18.5	14.8	12.4	10.6	9.2	7.4	6.2
10	7.48	0.16	233.3	15.0	12.0	10.0	8.6	7.5	6.0	5.0
11	6.78	0.19	256.7	12.3	9.9	8.2	7.0	6.2	4.9	4.1
12	6.19	0.23	280.0	10.3	8.3	6.9	5.9	5.2	4.1	3.4
13	5.69	0.27	303.3	8.8	7.0	5.8	5.0	4.4	3.5	2.9
14	5.26	0.31	326.7	7.5	6.0	5.0	4.3	3.7	3.0	2.5
15	4.89	0.35	350.0	6.5	5.2	4.3	3.7	3.2	2.7	2.2
16	4.56	0.40	373.3	5.7	4.6	3.8	3.2	2.8	2.4	1.9
17	4.27	0.46	396.7	5.0	4.0	3.3	2.9	2.5	2.0	1.7
18	4.01	0.51	420.0	4.5	3.6	3.0	2.6	2.2	1.8	1.5
19	3.78	0.57	443.3	4.0	3.2	2.6	2.3	2.0	1.6	1.3
20	3.57	0.63	466.7	3.6	2.9	2.4	2.1	1.8	1.4	1.2
21	3.37	0.70	490.0	3.2	2.6	2.1	1.8	1.6	1.3	1.1
22	3.20	0.77	513.3	2.9	2.3	1.9	1.6	1.4	1.2	1.0
23	3.04	0.84	536.7	2.6	2.1	1.8	1.5	1.3	1.0	0.9
24	2.89	0.91	560.0	2.4	1.9	1.6	1.3	1.2	1.0	0.8
25	2.75	0.99	583.3	2.2	1.7	1.5	1.2	1.1	0.9	0.7
26	2.62	1.07	606.7	2.0	1.6	1.3	1.1	1.0	0.8	0.7
27	2.50	1.16	630.0	1.8	1.4	1.2	1.0	0.9	0.7	0.6
28	2.39	1.26	653.3	1.7	1.4	1.1	1.0	0.8	0.7	0.6
29	2.28	1.33	676.7	1.6	1.3	1.0	0.9	0.8	0.6	0.5
30	2.18	1.43	700.0	1.5	1.2	1.0	0.8	0.7	0.6	0.5
31	2.08	1.53	723.3	1.3	1.1	0.9	0.7	0.7	0.5	0.4
32	1.99	1.63	746.7	1.2	1.0	0.8	0.7	0.6	0.5	0.4
33	1.91	1.74	770.0	1.2	1.0	0.8	0.7	0.6	0.5	0.4
34	1.83	1.84	793.3	1.1	0.9	0.7	0.6	0.5	0.4	0.4
35	1.76	1.95	816.7	1.1	0.9	0.7	0.6	0.5	0.4	0.4
36	1.69	2.06	840.0	0.9	0.7	0.6	0.5	0.5	0.4	0.3
37	1.62	2.17	863.3	0.9	0.7	0.6	0.5	0.4	0.4	0.3
38	1.55	2.29	886.7	0.8	0.7	0.5	0.5	0.4	0.3	0.3
39	1.49	2.41	910.0	0.8	0.7	0.5	0.5	0.4	0.3	0.3
40	1.43	2.54	933.3	0.7	0.6	0.5	0.4	0.3	0.3	0.2

9-INCH HEAVY BEAM—85 LBS. PER YARD.

Trenton Beam.		Length, in inches.	Weight, in lbs.	Length, in feet.	Weight, in lbs.
	Depth, 9 inches.	1	2.4	1	28.3
		2	4.7	2	56.7
	Width of flanges, 4 inches.	3	7.1	3	85.0
		4	9.4	4	113.3
		5	11.8	5	141.7
	Thickness of stem, .38 inches.	6	14.2	6	170.0
		7	16.5		
		8	18.9		
	Area of cross-section, 8.32 sq. inches.	9	21.2		
		10	23.6		
		11	26.0		

Distance between supports, in feet.	Safe uniformly distributed load, in tons of 2,000 lbs.	Deflection, in inches, under this load.	Weight in lbs.	Proper Distance, in feet, between Centres of Beams, for load, in lbs., per square foot of—						
				100 lbs.	125 lbs.	150 lbs.	175 lbs.	200 lbs.	250 lbs.	300 lbs.
6	11.76	0.05	170.0	39.2	31.4	26.1	22.4	19.6	15.7	13.1
7	11.73	0.07	198.3	33.5	26.8	22.3	19.1	16.7	13.4	11.2
8	11.70	0.10	226.7	29.2	25.4	19.5	16.7	14.6	11.7	9.7
9	10.37	0.13	255.0	23.0	18.4	15.4	13.1	11.5	9.2	7.7
10	9.31	0.16	283.3	18.6	14.9	12.4	10.6	9.3	7.4	6.2
11	8.43	0.19	311.7	15.3	12.3	10.2	8.7	7.6	6.1	5.1
12	7.70	0.23	340.0	12.8	10.3	8.6	7.3	6.4	5.1	4.3
13	7.08	0.27	368.3	10.9	8.7	7.3	6.2	5.4	4.4	3.6
14	6.55	0.31	396.7	9.4	7.5	6.2	5.3	4.7	3.7	3.1
15	6.09	0.35	425.0	8.1	6.5	5.4	4.6	4.0	3.2	2.7
16	5.68	0.40	453.3	7.1	5.7	4.7	4.1	3.5	2.8	2.4
17	5.32	0.46	481.7	6.3	5.0	4.2	3.6	3.1	2.5	2.1
18	5.00	0.51	510.0	5.5	4.4	3.7	3.2	2.7	2.2	1.8
19	4.70	0.57	538.3	4.9	3.9	3.3	2.8	2.4	2.0	1.6
20	4.44	0.63	566.7	4.4	3.5	3.0	2.5	2.2	1.8	1.5
21	4.20	0.70	595.0	4.0	3.2	2.7	2.3	2.0	1.6	1.3
22	3.98	0.77	623.3	3.6	2.9	2.4	2.1	1.8	1.4	1.2
23	3.78	0.84	651.7	3.3	2.7	2.2	1.9	1.6	1.3	1.1
24	3.60	0.91	680.0	3.0	2.4	2.0	1.7	1.5	1.2	1.0
25	3.43	0.99	708.3	2.7	2.2	1.8	1.5	1.3	1.1	0.9
26	3.27	1.07	736.7	2.5	2.0	1.7	1.4	1.2	1.0	0.8
27	3.13	1.16	765.0	2.3	1.8	1.5	1.3	1.1	0.9	0.8
28	2.98	1.24	793.3	2.1	1.7	1.4	1.2	1.0	0.9	0.7
29	2.85	1.33	821.7	2.0	1.6	1.3	1.1	1.0	0.8	0.7
30	2.72	1.43	850.0	1.8	1.5	1.2	1.0	0.9	0.7	0.6
31	2.61	1.53	878.3	1.7	1.4	1.1	1.0	0.8	0.7	0.6
32	2.50	1.63	906.7	1.6	1.3	1.0	0.9	0.8	0.6	0.5
33	2.40	1.74	935.0	1.5	1.2	1.0	0.9	0.7	0.6	0.5
34	2.30	1.84	963.3	1.4	1.1	0.9	0.8	0.7	0.6	0.5
35	2.20	1.95	991.7	1.3	1.0	0.8	0.7	0.6	0.5	0.4
36	2.11	2.06	1020.0	1.2	1.0	0.8	0.7	0.6	0.5	0.4
37	2.02	2.17	1048.3	1.1	0.9	0.7	0.6	0.5	0.4	0.4
38	1.94	2.29	1076.7	1.0	0.8	0.7	0.6	0.5	0.4	0.3
39	1.87	2.41	1105.0	1.0	0.8	0.6	0.6	0.5	0.4	0.3
40	1.80	2.54	1133.3	0.9	0.7	0.6	0.5	0.4	0.4	0.3

9-INCH EXTRA HEAVY BEAM—125 LBS. PER YARD.

TRENTON BEAM.	Length, in inches.	Weight, in lbs.	Length, in feet.	Weight, in lbs.
Depth, 9 inches.	1	3.5	1	41.7
	2	7.0	2	83.3
Width of flanges, 4½ inches.	3	10.4	3	125.0
	4	13.9	4	166.7
	5	17.4	5	208.3
Thickness of stem, .57 inch.	6	20.8	6	250.0
	7	24.3		
	8	27.8		
Area of cross-section, 12.33 sq. inch.	9	31.2		
	10	34.7		
	11	38.2		

Distance between supports in feet.	Safe uniformly distributed in load tons of 2,000 lbs.	Deflection, in inches, under this load.	Weight in lbs.	Proper Distance, in feet, between Centres of Beams, for load, in lbs., per square foot of—						
				100 lbs.	125 lbs.	150 lbs.	175 lbs.	200 lbs.	250 lbs.	300 lbs.
6	16.62	0.05	250.0	55.4	44.3	36.9	31.6	27.7	22.2	18.5
7	16.60	0.07	291.7	47.4	37.9	31.6	27.1	23.7	19.1	15.8
8	16.58	0.10	333.3	41.5	33.2	27.7	23.7	20.7	16.6	13.8
9	14.70	0.13	375.0	32.7	26.2	21.8	18.7	16.3	13.1	10.9
10	13.19	0.16	416.7	26.4	21.1	17.6	15.1	13.2	10.6	8.8
11	11.95	0.19	458.3	21.7	17.4	14.5	12.4	10 9	8.7	7.2
12	10.92	0.23	500.0	18.2	14.6	12.1	10.4	9.1	7.3	6.0
13	10.04	0.27	541.7	15.4	12.4	10.3	8.8	7.7	6.2	5.1
14	9.28	0.31	583.3	13.3	10.6	8.8	7.6	6.6	5.3	4.4
15	8.62	0.35	625.0	11.5	9.2	7.7	6.6	5.7	4.6	3.8
16	8.04	0.40	666.7	10.0	8.0	6.7	5.7	5.0	4.0	3.3
17	7.53	0.46	708.3	8.8	7.0	5.9	5.0	4.4	3.5	2.9
18	7.07	0.51	750.0	7.8	6.2	5.2	4.5	3.9	3.1	2.6
19	6.66	0.57	791.7	7.0	5.6	4.7	4.0	3.5	2.8	2.3
20	6.28	0.63	833.0	6.3	5.0	4.2	3.6	3.1	2.5	2.1
21	5.94	0.70	875.3	5.7	4.6	3.8	3.2	2.8	2.3	1.9
22	5.63	0.77	916.7	5.1	4.1	3.4	2.9	2.5	2.0	1.7
23	5.35	0.84	958.3	4.6	3.7	3.1	2.6	2.3	1.8	1.5
24	5.08	0.91	1000.0	4.2	3.4	2.8	2.4	2.1	1.7	1.4
25	4.84	0.99	1041.7	3.9	3.1	2.6	2.2	1.9	1.5	1.3
26	4.61	1.07	1083.3	3.5	2.8	2.4	2.0	1.8	1.4	1.2
27	4.40	1.16	1125.0	3.3	2.6	2.2	1.9	1.6	1.3	1.1
28	4.20	1.24	1166.7	3.0	2.4	2.0	1.7	1.5	1.2	1.0
29	4.02	1.33	1208.3	2.8	2.2	1.8	1.6	1.4	1.1	0.9
30	3.84	1.43	1250.0	2.6	2.1	1.7	1.5	1.3	1.0	0.9
31	3.68	1.53	1291.7	2.4	1.9	1.6	1.4	1.2	1.0	0.8
32	3.52	1.63	1333.3	2.2	1.8	1.5	1.3	1.1	0.9	0.7
33	3.37	1.74	1375.0	2.0	1.6	1.4	1.2	1.0	0.8	0.7
34	3.23	1.84	1416.7	1.9	1.5	1.3	1.1	0.9	0.8	0.6
35	3.10	1.95	1458.3	1.8	1.4	1.2	1.0	0.9	0.7	0.6
36	2.97	2.06	1500.0	1.6	1.3	1.1	0.9	0.8	0.7	0.5
37	2.85	2.17	1541.7	1.5	1.2	1.0	0.9	0.7	0.6	0.5
38	2.73	2.29	1583.3	1.4	1.1	1.0	0.8	0.7	0.6	0.5
39	2.62	2.41	1625.0	1.3	1.0	0.9	0.7	0 6	0.5	0.4
40	2.52	2.54	1666.7	1.3	1.0	0.8	0.7	0.6	0.5	0.4

10½-INCH LIGHT BEAM—105 LBS. PER YARD.

TRENTON BEAM.		Length, in inches.	Weight, in lbs.	Length, in feet.	Weight, in lbs.
Depth, 10¼ inches.		1	2.9	1	35
		2	5.8	2	70
Width of flanges, 4½ inches.		3	8.7	3	105
		4	11.7	4	140
		5	14.6	5	175
Thickness of stem, ⅝ inch.		6	17.5	6	210
		7	20.4		
		8	23.3		
Area of cross-section, 10.44 sq. inches.		9	26.2		
		10	29.2		
		11	32.1		

Distance between supports, in feet.	Safe uniformly distributed load, in tons of 2,000 lbs.	Deflection, in inch., under this load.	Weight, in lbs.	Proper Distance, in feet, between Centres of Beams, for load in lbs., per square foot of—						
				100 lbs.	125 lbs.	150 lbs.	175 lbs.	200 lbs.	250 lbs.	300 lbs.
6	12.90	0.03	210	43.0	34.4	28.7	24.6	21.5	17.2	14.3
7	12.88	0.05	245	36.8	29.4	24.5	21.0	18.4	14.7	12.3
8	12.86	0.07	280	32.1	25.7	21.4	18.3	16.0	12.8	10.7
9	12.84	0.10	315	28.5	22.8	19.0	16.3	14.2	11.4	9.6
10	12.83	0.14	350	25.7	20.6	17.1	14.7	12.8	10.3	8.6
11	12.81	0.16	385	23.3	18.6	15.5	13.3	11.6	9.3	7.8
12	11.71	0.19	420	19.5	15.6	13.0	11.1	9.7	7.8	6.5
13	10.77	0.23	455	16.6	13.3	11.0	9.5	8.3	6.6	5.5
14	9.97	0.27	490	14.2	11.4	9.5	8.1	7.1	5.7	4.7
15	9.27	0.31	525	12.4	9.9	8.2	7.1	6.2	5.0	4.1
16	8.66	0.35	560	10.8	8.6	7.2	6.2	5.4	4.3	3.6
17	8.11	0.39	595	9.6	7.7	6.4	5.5	4.8	3.8	3.2
18	7.63	0.44	630	8.5	6.8	5.6	4.9	4.2	3.4	2.8
19	7.19	0.49	665	7.6	6.1	5.0	4.3	3.8	3.0	2.5
20	6.80	0.54	700	6.8	5.4	4.5	3.9	3.4	2.7	2.3
21	6.44	0.60	735	6.1	4.9	4.1	3.5	3.0	2.4	2.0
22	6.11	0.66	770	5.6	4.5	3.7	3.2	2.8	2.2	1.9
23	5.81	0.72	805	5.1	4.1	3.4	2.9	2.5	2.0	1.7
24	5.54	0.78	840	4.6	3.7	3.1	2.6	2.3	1.8	1.5
25	5.28	0.85	875	4.2	3.4	2.8	2.4	2.1	1.7	1.4
26	5.04	0.92	910	3.9	3.1	2.6	2.2	1.9	1.6	1.3
27	4.82	0.99	945	3.6	2.9	2.4	2.1	1.8	1.4	1.2
28	4.62	1.07	980	3.3	2.7	2.2	1.9	1.6	1.3	1.1
29	4.42	1.14	1015	3.0	2.4	2.0	1.7	1.5	1.2	1.0
30	4.24	1.22	1050	2.8	2.2	1.9	1.6	1.4	1.1	0.9
31	4.07	1.30	1085	2.6	2.1	1.7	1.5	1.3	1.0	0.9
32	3.91	1.39	1120	2.4	1.9	1.6	1.4	1.2	0.9	0.8
33	3.75	1.48	1155	2.3	1.8	1.5	1.3	1.1	0.9	0.8
34	3.61	1.57	1190	2.1	1.7	1.4	1.2	1.0	0.8	0.7
35	3.47	1.67	1225	2.0	1.6	1.3	1.1	1.0	0.8	0.7
36	3.34	1.76	1260	1.9	1.5	1.2	1.1	0.9	0.8	0.6
37	3.21	1.86	1295	1.7	1.4	1.2	1.0	0.8	0.7	0.6
38	3.09	1.96	1330	1.6	1.3	1.1	1.0	0.8	0.6	0.5
39	2.98	2.07	1365	1.5	1.2	1.0	0.9	0.7	0.6	0.5
40	2.87	2.18	1400	1.4	1.1	1.0	0.8	0.7	0.6	0.5

10½-INCH HEAVY BEAM—135 LBS. PER YARD.

Trenton Beam.	Length, in inches.	Weight, in lbs.	Length, in feet.	Weight, in lbs.
Depth, 10½ inches.	1	3.7	1	45
	2	7.5	2	90
Width of flanges, 5 inches.	3	11.2	3	135
	4	15.0	4	180
	5	18.7	5	225
Thickness of stem, .47 inch.	6	22.5	6	270
	7	26.2		
	8	30.0		
Area of cross-section, 13.36 sq. inches.	9	33.7		
	10	37.5		
	11	41.2		

Distance between supports, in feet.	Safe uniformly distributed, load in tons of 2,000 lbs.	Deflection, in inches, under this load.	Weight, in lbs.	Proper Distance in feet, between Centres of Beams, for load, in lbs., per square foot of—						
				100 lbs.	125 lbs.	150 lbs.	175 lbs.	200 lbs.	250 lbs.	300 lbs.
6	16.22	0.03	270	54.1	43.3	36.0	30.9	27.0	21.6	18.0
7	16.20	0.05	315	46.3	37.0	30.9	26.5	23.1	18.5	15.4
8	16.18	0.07	360	40.4	32.3	27.0	23.1	20.2	16.2	13.5
9	16.16	0.10	405	35.9	28.7	24.0	20.5	18.0	14.4	12.0
10	16.14	0.14	450	32.3	25.8	21.5	18.5	16.1	12.9	10.8
11	16.12	0.16	495	29.3	23.4	19.5	16.7	14.6	11.7	9.8
12	14.73	0.19	540	24.5	19.6	16.4	14.0	12.2	9.8	8.2
13	13.55	0.23	585	20.8	16.7	13.9	11.9	10.4	8 3	6.9
14	12.54	0.27	630	17.9	14.3	11.9	10.2	8.9	7.2	6.0
15	11.66	0.31	675	15.5	12.4	10.4	8.8	7.7	6.2	5.2
16	10 89	0.35	720	13.6	10.9	9.1	7.8	6.8	5 4	4.5
17	10.20	0.39	765	12.0	9.6	8.0	6.9	6.0	4.8	4 0
18	9.59	0.44	810	10.7	8.5	7.1	6.1	5.3	4.3	3.6
19	9.05	0.49	855	9.5	7.6	6.3	5.4	4.7	3.8	3.2
20	8.55	0.54	900	8.6	6.9	5.7	4.9	4.3	3.4	2.9
21	8.10	0.60	945	7.7	6.2	5.1	4.4	3.8	3.1	2.6
22	7.69	0.66	990	7.0	5.6	4.7	4.0	3.5	2.8	2.3
23	7.31	0.72	1035	6.4	5.1	4.2	3.7	3.2	2.6	2.1
24	6.96	0.78	1080	5.8	4.6	3.9	3.3	2.9	2.3	1.9
25	6.64	0.85	1125	5.3	4.2	3.5	3.0	2.6	2.1	1.8
26	6.34	0.92	1170	4.9	3.9	3.3	2.8	2.4	2.0	1.6
27	6.06	0.99	1215	4.5	3.6	3.0	2.6	2.2	1.8	1.5
28	5.80	1.07	1260	4.1	3.3	2.8	2.4	2.0	1.6	1.4
29	5.55	1.14	1305	3.8	3.0	2.6	2.2	1.9	1.5	1.3
30	5.32	1.22	1350	3.5	2.8	2.4	2.0	1.7	1.4	1.2
31	5.11	1.30	1395	3.3	2.6	2.2	1.9	1.6	1.3	1 1
32	4.90	1.39	1440	3.1	2.5	2.0	1.8	1.5	1.2	1.0
33	4.71	1.48	1485	2.9	2.3	1.9	1.7	1.4	1.2	1.0
34	4.53	1.57	1530	2.7	2.2	1.8	1.6	1.3	1.1	0.9
35	4.35	1.67	1575	2.5	2.0	1.7	1.4	1.2	1.0	0.8
36	4.19	1.76	1620	2.3	1.8	1.6	1.3	1.1	0.9	0.8
37	4.03	1.86	1665	2.2	1.7	1.5	1.2	1.1	0 9	0.7
38	3.88	1.96	1710	2.0	1.6	1.4	1.0	1.0	0.8	0.7
39	3.74	2.07	1755	1.9	1.5	1.3	1.0	0.9	0.8	0.6
40	3.60	2.18	1800	1.8	1.4	1.2	1.0	0.9	0.7	0.6

12¼-INCH LIGHT BEAM—125 LBS. PER YARD.

TRENTON BEAM.		Length, in inches.	Weight, in lbs.	Length, in feet.	Weight, in lbs.
	Depth, 12¼ inches.	1	3.5	1	41.7
		2	6.9	2	83.3
	Width of flanges, 4.8 inches.	3	10.4	3	125.0
		4	13.9	4	166.7
		5	17.4	5	208.3
	Thickness of stem, 0.47 inch.	6	20 8	6	250.0
		7	24.3		
		8	27.8		
	Area of cross-section, 12.33 sq. inches.	9	31.2		
		10	34.7		
		11	38.2		

Distance between supports, in feet.	Safe uniformly distributed load in tons of 2,000 lbs.	Deflection, in inches, under this load.	Weight, in lbs.	Proper Distance, in feet, between Centres of Beams, for load, in lbs., per square foot of—						
				100 lbs.	125 lbs.	150 lbs.	175 lbs.	200 lbs.	250 lbs.	300 lbs.
6	18.73	0.03	250.0	62.4	49.9	41.6	35.7	31.2	25.0	20.8
7	18.70	0.05	291.7	53.4	42.7	35.6	30.5	26.7	21.4	17.8
8	18 68	0.07	333.3	46.7	37.4	31.2	26.7	23 3	18.7	15.6
9	18.66	0.09	375.0	41.5	33.2	27.6	23.7	20.7	16.6	13.8
10	18.64	0.12	416.7	37.3	29.8	24.8	21.3	18.6	14.9	12.4
11	16.91	0.14	458.3	30.7	24 6	20.5	17.5	15.3	12.3	10.2
12	15.46	0.17	500.0	25 8	20.6	17.2	14.7	12.9	10.3	8.6
13	14.23	0.20	541 7	21.9	17.5	14.6	12.5	10.9	8.8	7.3
14	13.17	0.23	583.3	18.8	15.0	12.5	10.7	9.4	7.5	6.3
15	12.25	0.26	625.0	16.3	13.0	10.9	9.3	8.1	6.5	5.4
16	11.45	0 30	666.7	14.3	11.4	9.5	8.2	7.1	5.7	4.8
17	10.73	0.34	708.3	12.6	10.1	8.4	7.2	6.3	5.0	4.2
18	10.10	0.38	750.0	11.2	9.0	7.5	6.4	5.6	4.5	3.7
19	9.52	0.42	791.7	10.0	8.0	6.7	5.7	5.0	4.0	3.3
20	9.01	0.46	833.3	9.0	7.2	6.0	5.1	4.5	3.6	3.0
21	8.54	0.51	875.0	8.1	6 5	5.4	4.6	4.0	3.2	2.7
22	8.11	0.56	916.7	7.4	5.9	4.9	4.2	3.7	3.0	2.5
23	7.72	0.62	958.3	6.7	5.4	4.5	3.8	3.3	2.7	2.2
24	7.35	0.67	1000.0	6.1	4.9	4.1	3.5	3.0	2.4	2 0
25	7.02	0.73	1041.7	5.6	4.5	3.7	3.2	2.8	2.2	1.9
26	6.71	0.79	1083.3	5.1	4 1	3.4	2.9	2.5	2.0	1.7
27	6.42	0.85	1125.0	4.7	3.7	3.2	2.7	2.3	1.9	1.6
28	6.15	0.91	1166.7	4.4	3.5	2.9	2.5	2.2	1.8	1.5
29	5.90	0.98	1208.3	4.1	3.3	2.7	2.3	2.0	1.6	1.4
30	5.66	1.05	1250.0	3.8	3.1	2.5	2.2	1 9	1.5	1.3
31	5.43	1.12	1291.7	3 5	2.8	2.3	2.0	1.7	1.4	1.2
32	5.22	1.10	1333.3	3.3	2.6	2.2	1.9	1.6	1.3	1.1
33	5.02	1.27	1375.0	3.1	2.5	2.0	1.8	1.5	1.2	1.0
34	4.83	1.35	1416.7	2.8	2.3	1.9	1.6	1.4	1.1	0.9
35	4.66	1 43	1458.3	2.7	2.1	1.8	1.5	1.3	1.1	0.9
36	4.49	1.52	1500.0	2.5	2.0	1.7	1.4	1.2	1.0	0.8
37	4.32	1 60	1541.7	2.3	1.9	1.6	1.3	1.1	0.9	0.8
38	4.17	1.68	1583.3	2.2	1.8	1.5	1.2	1.1	0.9	0.7
39	4.02	1.77	1625.0	2.1	1.7	1.4	1.2	1.0	0.8	0.7
40	3.88	1.87	1666.7	1.9	1.5	1.3	1.1	0.9	0.8	0.6

12¼-INCH HEAVY BEAM—170 LBS. PER YARD.

TRENTON BEAM.		Length, in inches.	Weight, in lbs.	Length, in feet.	Weight, in lbs.
Depth, 12¹⁵⁄₁₆ inches.		1	4.7	1	56.7
		2	9.4	2	113.3
Width of flanges, 5¼ inches.		3	14.2	3	170.0
		4	18.9	4	226.7
		5	23.6	5	283.3
Thickness of stem, 0.6 inch.		6	28.3	6	340.0
		7	33.1		
		8	37.8		
Area of cross-section, 16.77 sq. inches.		9	42.5		
		10	47.2		
		11	51.9		

Distance between supports, in feet.	Safe uniformly distributed load in tons of 2,000 lbs.	Deflection, in inches, under this load.	Weight, in lbs.	Proper Distance, in feet, between Centres of Beams, for load, in lbs., per square foot of—						
				100 lbs.	125 lbs.	150 lbs.	175 lbs.	200 lbs.	250 lbs.	300 lbs.
6	25.39	0.03	340.0	84.6	67.8	56.4	48.3	42.3	33.8	28.2
7	25.36	0.05	396.7	72.5	58.0	48.3	41.4	36.2	29.0	24.2
8	25.33	0.07	453.3	63.3	50.6	42.2	36.2	31.6	25.3	21.1
9	25.30	0.09	510.0	56.2	44.9	37.5	32.1	28.1	22.5	18.7
10	25.27	0.12	566.7	50.5	40.4	33.7	28.9	25.2	20.2	16.8
11	22.91	0.14	623.3	41.6	33.3	27.7	23.8	20.8	16.6	13.9
12	20.95	0.16	680.0	34.9	27.9	23.3	19.9	17.4	13.9	11.6
13	19.28	0.20	736.7	29.7	23.7	19.8	17.0	14.8	11.9	9.9
14	17.85	0.23	793.3	25.5	20.4	17.0	14.6	12.7	10.2	8.5
15	16.61	0.26	850.0	22.1	17.7	14.8	12.6	11.0	8.9	7.4
16	15.52	0.30	906.7	19.4	15.5	12.9	11.1	9.7	7.8	6.5
17	14.55	0.34	963.3	17.1	13.7	11.4	9.7	8.5	7.0	5.7
18	13.68	0.38	1020.0	15.2	12.1	10.1	8.7	7.6	6.1	5.1
19	12.91	0.42	1076.7	13.6	10.9	9.0	7.8	6.8	5.4	4.5
20	12.21	0.46	1133.3	12.2	9.9	8.1	7.0	6.1	4.9	4.1
21	11.57	0.51	1190.0	11.0	8.9	7.3	6.3	5.5	4.4	3.7
22	10.99	0.56	1246.7	10.0	8.0	6.7	5.7	5.0	4.0	3.3
23	10.46	0.62	1303.3	9.1	7.3	6.1	5.2	4.5	3.6	3.0
24	9.96	0.67	1360.0	8.3	6.7	5.5	4.7	4.1	3.3	2.8
25	9.51	0.73	1416.7	7.6	6.1	5.1	4.3	3.8	3.0	2.5
26	9.09	0.79	1473.3	7.0	5.6	4.7	4.0	3.5	2.8	2.3
27	8.70	0.84	1530.0	6.4	5.1	4.3	3.7	3.2	2.6	2.1
28	8.33	0.91	1586.7	5.9	4.7	4.0	3.4	2.9	2.4	2.0
29	7.99	0.98	1643.3	5.5	4.4	3.7	3.1	2.7	2.2	1.8
30	7.67	1.05	1700.0	5.1	4.1	3.4	2.9	2.5	2.0	1.7
31	7.36	1.12	1756.7	4.7	3.7	3.2	2.7	2.3	1.9	1.6
32	7.08	1.20	1813.3	4.4	3.5	2.9	2.5	2.2	1.8	1.5
33	6.81	1.27	1870.0	4.1	3.3	2.7	2.3	2.0	1.6	1.4
34	6.55	1.35	1926.7	3.8	3.0	2.6	2.2	1.9	1.5	1.3
35	6.31	1.43	1983.3	3.6	2.9	2.4	2.1	1.8	1.4	1.2
36	6.08	1.52	2040.0	3.4	2.7	2.2	1.9	1.7	1.4	1.1
37	5.86	1.60	2096.7	3.2	2.6	2.1	1.8	1.6	1.3	1.1
38	5.65	1.68	2153.3	3.0	2.4	2.0	1.7	1.5	1.2	1.0
39	5.45	1.77	2210.0	2.8	2.3	1.9	1.6	1.4	1.1	0.9
40	5.25	1.87	2266.7	2.6	2.1	1.7	1.5	1.3	1.0	0.9

15-INCH LIGHT BEAM—150 LBS. PER YARD.

Trenton Beam.		Length, in inches.	Weight, in lbs.	Length, in feet.	Weight, in lbs.
Depth, 15 3/16 inches.		1	4.2	1	50
		2	8.3	2	100
Width of flanges, 5 inches.		3	12.5	3	150
		4	16.7	4	200
		5	20.8	5	250
Thickness of stem, 0.5 inch.		6	25.0	6	300
		7	29.2		
		8	33.3		
Area of cross-section, 15.04 sq. inches.		9	37.5		
		10	41.7		
		11	45.8		

Distance between supports, in feet.	Safe uniformly distributed load, in tons of 2,000 lbs.	Deflection, in inches, under this load.	Weight, in lbs.	Proper Distance, in feet, between Centres of Beams, for load, in lbs., per square foot of—						
				100 lbs.	125 lbs.	150 lbs.	175 lbs.	200 lbs.	250 lbs.	300 lbs.
6	24.90	0.02	300	83.0	66.4	55.3	47.4	41.5	33.2	27.7
7	24.88	0.03	350	71.1	56.9	47.4	40.6	35.5	28.4	23.7
8	24.85	0.05	400	62.1	49.7	41.4	35.5	31.1	24.8	20.7
9	24.83	0.07	450	55.2	44.1	36.8	31.5	27.6	22.1	18.4
10	24.80	0.09	500	49.6	39.7	33.1	28.3	24.8	19.8	16.5
11	24.77	0.12	550	45.0	36.0	30.0	25.7	22.5	18.0	15.0
12	22.66	0.14	600	37.8	30.2	25.2	21.6	18.9	15.1	12.6
13	20.87	0.16	650	32.1	25.7	21.4	18.3	16.0	12.8	10.7
14	19.33	0.19	700	27.6	23.1	18.4	15.8	13.8	11.0	9.2
15	17.99	0.21	750	24.0	19.2	16.0	13.7	12.0	9.6	8.0
16	16.82	0.24	800	21.0	16.8	14.0	12.0	10.5	8.4	7.0
17	15.78	0.28	850	18.6	14.9	12.4	10.6	9.3	7.4	6.2
18	14.85	0.31	900	16.5	13.2	11.0	9.4	8.3	6.6	5.5
19	14.02	0.34	950	14.8	11.8	9.8	8.4	7.4	5.9	4.9
20	13.27	0.38	1000	13.3	10.6	8.8	7.6	6.6	5.3	4.4
21	12.59	0.42	1050	12.0	9.6	8.0	6.9	6.0	4.8	4.0
22	11.97	0.46	1100	10.9	8.7	7.2	6.2	5.4	4.3	3.6
23	11.40	0.50	1150	9.9	7.9	6.6	5.6	4.9	3.9	3.3
24	10.88	0.55	1200	9.1	7.3	6.0	5.2	4.5	3.6	3.0
25	10.39	0.59	1250	8.3	6.7	5.5	4.7	4.1	3.3	2.8
26	9.95	0.64	1300	7.6	6.1	5.1	4.3	3.8	3.0	2.5
27	9.53	0.69	1350	7.1	5.6	4.7	4.0	3.5	2.8	2.4
28	9.14	0.75	1400	6.5	5.2	4.3	3.7	3.2	2.6	2.3
29	8.77	0.80	1450	6.0	4.8	4.0	3.4	3.0	2.4	2.0
30	8.43	0.86	1500	5.6	4.5	3.7	3.2	2.8	2.2	1.9
31	8.11	0.92	1550	5.2	4.2	3.5	3.0	2.6	2.1	1.7
32	7.81	0.98	1600	4.9	3.9	3.2	2.8	2.4	1.9	1.6
33	7.52	1.04	1650	4.5	3.6	3.0	2.6	2.2	1.8	1.5
34	7.25	1.10	1700	4.3	3.4	2.8	2.5	2.1	1.7	1.4
35	7.00	1.17	1750	4.0	3.2	2.7	2.3	2.0	1.6	1.3
36	6.75	1.23	1800	3.7	3.0	2.5	2.1	1.8	1.5	1.2
37	6.52	1.30	1850	3.5	2.8	2.3	2.0	1.7	1.4	1.2
38	6.30	1.37	1900	3.3	2.6	2.2	1.9	1.6	1.3	1.1
39	6.09	1.45	1950	3.1	2.5	2.1	1.8	1.5	1.2	1.0
40	5.89	1.52	2000	2.9	2.3	2.0	1.7	1.4	1.2	1.0

15-INCH HEAVY BEAM—200 LBS. PER YARD.

TRENTON BEAM.	Length, in inches.	Weight, in lbs.	Length, in feet.	Weight, in lbs.
Depth, 15¼ inches.	1	5.6	1	66.7
	2	11.1	2	133.3
Width of flanges, 5¼ inches.	3	16.7	3	200.0
	4	22.2	4	266.7
	5	27.7	5	333.3
Thickness of stem, 0.6 inch.	6	33.3		
	7	38.9		
	8	44.4		
Area of cross-section, 20.02 sq. inches.	9	50.0		
	10	55.6		
	11	61.1		

Distance between supports, in feet.	Safe uniformly distributed load in tons of 2,000 lbs.	Deflection, in inches, under this load.	Weight, in lbs.	Proper Distance, in feet, between Centres of Beams, for load, in lbs., per square foot of—						
				100 lbs.	125 lbs.	150 lbs.	175 lbs.	200 lbs.	250 lbs.	300 lbs.
6	30.97	0.02	400.0	103.2	82.6	68.8	59.0	51.6	41.4	34.4
7	30.93	0.03	466.7	88.4	70.7	58.9	50.5	44.2	35.	29.5
8	30.90	0.05	533.3	77.2	61.8	51.5	44.1	38.6	30.9	25.8
9	30.87	0.07	600.0	68.6	54.9	45.7	39.2	34.3	27.4	22.9
10	30.84	0.09	666.7	61.7	49.3	41.1	35.2	30.8	24.7	20.6
11	30.80	0.11	733.3	56.0	44.8	37.3	32.0	28.0	22.4	18.7
12	30.77	0.14	800.0	51.3	41.0	34.2	29.3	25.6	20.5	17.1
13	28.34	0.16	866.7	46.6	34.9	29.1	24.9	21.8	17.4	14.5
14	26.25	0.19	933.3	37.5	30.0	25.0	21.4	18.7	15.0	12.5
15	24.43	0.21	1000.0	32.6	26.1	21.7	18.6	16.3	13.0	10.9
16	22.84	0.24	1066.7	28.6	22.8	19.0	16.3	14.3	11.4	9.5
17	21.43	0.28	1133.3	25.2	20.2	16.8	14.4	12.6	10.1	8.4
18	20.18	0.31	1200.0	22.4	17.9	14.9	12.8	11.2	9.0	7.5
19	19.05	0.34	1266.7	20.0	16.0	13.4	11.5	10.0	8.0	6.7
20	18.03	0.38	1333.3	18.0	14.4	12.0	10.3	9.0	7.2	6.0
21	17.11	0.42	1400.0	16.3	13.0	10.9	9.3	8.1	6.5	5.4
22	16.27	0.46	1466.7	14.8	11.8	9.9	8.5	7.4	5.9	4.9
23	15.49	0.50	1533.3	13.5	10.8	9.0	7.7	6.7	5.4	4.5
24	14.78	0.55	1600.0	12.3	9.8	8.2	7.0	6.1	4.9	4.1
25	14.13	0.59	1666.7	11.3	9.0	7.5	6.5	5.6	4.5	3.8
26	13.52	0.64	1733.3	10.4	8.3	6.9	6.0	5.2	4.2	3.5
27	12.95	0.69	1800.0	9.6	7.7	6.4	5.5	4.8	3.8	3.2
28	12.42	0.75	1866.7	8.9	7.1	5.9	5.0	4.4	3.5	3.0
29	11.93	0.80	1933.3	8.2	6.6	5.5	4.7	4.1	3.3	2.7
30	11.47	0.86	2000.0	7.6	6.1	5.1	4.3	3.8	3.0	2.5
31	11.03	0.92	2066.7	7.1	5.7	4.7	4.0	3.5	2.8	2.4
32	10.62	0.98	2133.3	6.6	5.3	4.4	3.8	3.3	2.6	2.2
33	10.23	1.04	2200.0	6.2	5.0	4.1	3.5	3.1	2.5	2.1
34	9.87	1.10	2266.7	5.8	4.7	3.9	3.3	2.9	2.3	1.9
35	9.52	1.17	2333.3	5.4	4.3	3.6	3.1	2.7	2.2	1.8
36	9.19	1.23	2400.0	5.1	4.1	3.4	2.9	2.5	2.0	1.7
37	8.87	1.30	2466.7	4.8	3.8	3.2	2.7	2.4	1.9	1.6
38	8.57	1.37	2533.3	4.5	3.6	3.0	2.6	2.3	1.8	1.5
39	8.29	1.45	2600.0	4.3	3.4	2.8	2.5	2.2	1.7	1.4
40	8.02	1.52	2666.7	4.0	3.2	2.7	2.3	2.0	1.6	1.3

RELATIVE EFFICIENCY OF BEAM.

It is obviously most economical to employ the beam which shall support the greatest load in proportion to its weight. The following table presents the strength of each pattern of beam divided by its weight, and gives the means of comparing the economical efficiency of the different sections. The higher prices per pound of the larger beams diminish somewhat their relative economy; but still it will be found that, when the circumstances admit of their use, the deepest beams are the most economical.

TRENTON BEAM.	$\frac{c}{w} =$	TRENTON BEAM.	$\frac{c}{w} =$
4 inch, Light...............	10.03	9 inch, Light.............	21.73
" Heavy............	9.95	" Heavy............	22.28
5 inch, Light.............	12.90	" Extra...............	21.44
" Heavy............	12.27	10½ inch, Light............	27.20
6 inch, Light.............	15.65	" Heavy..........	26.64
" Heavy............	15.36	12¼ inch, Light...........	30.64
7 inch....................	16.92	" Heavy..........	28.41
8 inch, Light..............	20.75	15 inch, Light............	36.76
" Heavy............	20.99	" Heavy..........	37.41

SETTING AND CONNECTING BEAMS.

Beams for floors with brick arches should have a bearing on wall of about eight inches.

Tie-rods from ⅝ inch to ¾ inch diameter are ordinarily employed to take the thrust of the brick arches, and to add to the security of the floor. These may be spaced from eight to ten times the depth of the beam apart, and the holes for them are usually punched at the centre of the depth of the beam.

When beams are used to support walls, or as girders to carry floor-beams, they are often placed side by side, and should in this case be furnished with cast-iron separators fitting between the flanges, so as to firmly combine the two beams. These separators may be placed about the same distance apart as the tie-rods.

BEAMS UNSUPPORTED SIDEWAYS.

The foregoing tables are calculated on the assumption that the beams are secured against deflection sideways by filling in between them with brick arches, or in any other suitable manner. Beams unsupported sideways, of any considerable length, are liable to fall under a much lighter load by yielding laterally. The following table gives a comparison of the loads which will be supported safely in either case for each five feet of span:

Span, in feet.	TRENTON BEAM.	4 inch Light.	4 inch Heavy.	5 inch Light.	5 inch Heavy.	6 inch Light.	6 inch Heavy.	7 inch.	8 inch Light.	8 inch Heavy.
10	Supported........	1.45	1.78	1.88	2.40	3.06	3.76	4.95	6.64	8.27
	Unsupported.....	0.98	1 27	1.28	1.71	2.14	2.85	3.74	5.35	7.01
15	Supported........	0.93	1.13	1.22	1.54	1.99	2.44	3.22	4 34	5.40
	Unsupported.....	0.42	0.57	0.56	0.78	0.97	1.39	1.82	2.78	3.80
20	Supported........	0.65	0.80	0.87	1.10	1.43	1.75	2.32	3.16	3.93
	Unsupported.....	0.16	0.24	0.24	0.35	0.45	0.69	0.92	1.52	2.20
25	Supported........								2.43	3.03
	Unsupported.....								0.82	1.27

Span, in feet.	TRENTON BEAM.	9 inch Light.	9 inch Heavy.	9 inch Extra Heavy.	10½ inch Light.	10½ inch Heavy.	12¼ inch Light.	12¼ inch Heavy.	15 inch Light.	15 inch Heavy.
10	Supported........	7.48	9.31	13.19	12.83	16.13	18.64	25.27	24.80	30.87
	Unsupported.....	5.71	7.54	11.08	11.94	15.42	16.12	22.53	23.88	
15	Supported........	4.89	6.09	8.62	9.27	11.66	12.26	16.61	18.00	24.43
	Unsupported.....	2.84	3.94	5.98	6.53	8.63	9.00	13.00	13.55	19.37
20	Supported.......	3.57	4.44	6.28	6.80	8.55	9.01	12.20	13.27	18.03
	Unsupported.....	1.48	2.18	3.42	3.80	5.18	5.40	8.07	8.20	12.20
25	Supported........	2.75	3.43	4.84	5.28	6.44	7.02	9.51	10.40	14.13
	Unsupported.....	0.76	1.20	1.95	2.26	3.16	3.31	5.14	5.22	7 92

The rule by which this table is calculated will, of course, apply to beams of any span, and is as follows:

RULE.—Multiply the co-efficient for strength in column II. of the table of "Weights and Co-efficients" by the number given in column IV., headed "Correction for Lateral Deflection," and divide the product by the number in column IV., plus the square of the span taken in feet; this quotient divided by the span in feet will give the safe load in pounds.

WEIGHTS AND CO-EFFICIENTS FOR TRENTON ROLLED I BEAMS.

	I.	II.	III.	IV.	V.	VI.	VII.	VIII.	IX.	X.	XI.
					Strength as Strut.		Areas, in square inches.			Moments of Inertia.	
	Weight per yard, in lbs.	Co-efficient for Strength.	Maximum load, in lbs.	Correction for Lateral Deflection.	Sideways.	Edgeways.	Flanges.	Web.	Total Area.	Axis perpendicular to Web at Centre.	Axis co-incident with Centre Line of Web.
4 inch Light	30	30 100	7 250	215	96	644	1.91	1.00	2.91	7.5	1.11
4 " Heavy	37	36 800	9 200	263	121	634	2.41	1.25	3.66	9.2	1.74
5 " Light	30	38 700	9 670	218	87	996	1.79	1.25	3.04	12.1	1.04
5 " Heavy	40	49 100	9 820	261	109	996	2.34	1.56	3.90	15.4	1.68
6 " Light	40	62 600	10 400	239	102	1452	2.51	1.50	4.01	23.5	1.61
6 " Heavy	50	76 800	12 800	324	141	1475	3.11	1.80	4.91	29.0	2.74
7 "	60	102 000	16 900	317	132	1902	3.25	2.59	5.84	44.5	3.05
8 " Light	65	135 000	16 900	423	180	2632	3.97	2.40	6.37	67.4	4.55
8 " Heavy	60	168 000	18 700	566	240	2602	5.07	2.96	8.03	83.9	7.55
9 " Light	70	152 000	19 000	329	136	3277	3.83	2.70	6.53	85.6	3.50
9 " Heavy	85	189 000	23 700	435	170	3187	4.86	3.46	8.32	106.5	5.59
9 " Extra	125	268 000	33 500	535	227	3015	7.20	5.22	12.42	150.8	11.23
10½ " Light	105	286 000	26 000	555	230	4445	6.51	3.93	10.44	185.6	9.43
10½ " Heavy	135	360 000	32 700	663	300	4372	8.43	4.93	13.36	233.7	15.80
12½ " Light	125	377 000	37 700	645	235	5777	6.58	5.88	12.46	288.0	11.54
12½ " Heavy	170	511 000	51 100	835	382	5832	9.38	7.39	16.77	391.2	25.41
15 " Light	150	551 000	50 100	705	256	8702	7.45	7.59	15.04	523.5	15.29
15 " Heavy	200	748 000	62 400	882	343	8830	10.95	9.07	20.02	707.1	27.46

WEIGHTS AND CO-EFFICIENTS FOR TRENTON CHANNELS.

	I.	II.	IV.	V.	VI.	IX.	X.	XI.
	Weight per Yard. lbs.	Co-efficient for transverse strength.	Addition to Co-efficient for Increase of 1 lb. per foot in Weight of Bar.	Strength as Strut.		Area of Cross-Section, in square inches.	Moments of Inertia about Neutral Axis.	
				Sideways.	Edgeways.		Vertically.	Horizontally.
6 inch Light Channel	33	45,700	2,400	101	1,343	3.20	17.2	1.30
6 " Heavy "	45	58,300	2,400	123	1,257	4.32	21.7	2.12
9 " Light "	50	104,000	3,600	124	2,892	5.08	58.8	2.53
9 " Heavy "	70	146,000	3,600	190	2,925	7.02	82.1	5.35
12¼ " Light "	85	238,000	4,900	172	5,275	8.62	181.9	5.93
12¼ " Heavy "	140	381,000	4,900	317	5,170	14.10	291.6	17.87

BEAMS USED AS PILLARS OR STRUTS.

When a beam is used as a pillar or strut, and not as a girder, to find the safe load in tons of 2,000 lbs. which it will support. If secured against deflection, either by having accurately-faced capital and base, or in some other manner.

RULE.—Multiply the area of cross-section of the beam given in column IX. of the table of "Weights and Co-efficients" by 3, and multiply that product by the number given in column V., and divide the product so found by the number given in column V., plus the square of the longest length of the strut or pillar, which is unsupported sideways, taken in feet; or if, by reason of the pillar being supported sideways, it will fail, if at all, by deflection edgeways, substitute in the above rule for the number given in column V. that given in column VI., and for the longest length unsupported SIDEWAYS substitute the longest length unsupported EDGEWAYS.

If the pillar is HINGED or NOT FACED AT THE ENDS, and thus not secured against deflection, take in the foregoing rule one-fourth of the number in column V. or VI., instead of the whole number.

EXAMPLE.—What load will an 8-inch light beam, fifteen feet long, and having ends accurately faced, support as a pillar?

Area of cross-section, Col. IX. = 6.37.

Number in Col. V. = 180.

$$\frac{6.37 \times 3 \times 180}{180 + 225} = \frac{3440}{405} = 8.5 \text{ tons.}$$

If the strut is hinged at the ends so that its bearing opposes no resistance to deflection *sideways*, then we should use the number $\frac{180}{4} = 45$, instead of 180, and we should have for the load:

$$\frac{6.37 \times 3 \times 45}{45 + 225} = 3.2 \text{ tons.}$$

But if hinged so that it would deflect *edgeways* we should

use the number in column VI., divided by 4, viz. : $\dfrac{2632}{4} = 658$,

and the load would be

$$\frac{6.37 \times 3 \times 658}{658 + 225} = 14.2 \text{ tons,}$$

which is greater than the strength of the strut to resist deflection sideways, even when not hinged in that direction. Unless supported sideways, therefore, the load for such a strut would have to be limited to that found in the case first supposed, viz., 8.5 tons.

NOTES FOR ENGINEERS.

BASIS OF STRENGTH.

The co-efficients in the foregoing tables, except those in column III., headed "Maximum Load," correspond to a stress or straining force of 12,000 lbs. per square inch on the part of the beam at which the strain is a maximum. The co-efficient for strength, column II., divided by the clear span in feet, gives the safe uniformly distributed load of the beam in lbs. The greatest SHEARING STRESS on the stem under the loads, given in column III. as the maximum allowable, will be 4,000 lbs. per square inch. For any stress not exceeding the "limit of elasticity," which is about 21,000 lbs. per square inch, the amount of deflection will be in a certain direct proportion to the load applied, and on the removal of the load the beam will regain its original condition. For greater stresses the deflections will increase in a much more rapid ratio, and the beams will retain a "permanent set." Experiments on the effect of repeated applications and removals of the load, accompanied with considerable vibration, appear, however, to show that when a beam may be subjected to such repeated applications of the load an indefinitely great number of times, the maximum stress should not exceed 16,000 lbs. per square inch.

The basis adopted in the above table is therefore about one-quarter of the ultimate stress for a single application of the load ; four-sevenths of the limit of elasticity ; and three-quarters of the safe stress for indefinitely repeated applications of the load. The loads determined by the use of the co-efficients will therefore be the SAFE WORKING PERMANENT OR DEAD LOADS, including a sufficient margin of safe strength to allow for the vibrations and ordinary contingencies to which the floor-beams of buildings are subjected.

OTHER BASES OF STRENGTH.

If a greater or less basis of strength is preferred, the co-efficients corresponding to it are found by increasing or diminishing those given in columns II., IV., V., VI., in the same ratio as the basis assumed is greater or less than the basis of 12,000 lbs. taken for the table. The deflections will of course vary in the same ratio as the co-efficients.

CONDENSED TABLE OF WEIGHTS AND STRENGTHS.

Designation of Beam.	BEAMS. Weight per Foot in Pounds.	Co-efficient for Strength.
Light 4 inch.	10	30 100
Heavy 4 "	12⅓	36 800
Light 5 "	10	38 700
Heavy 5 "	13⅓	49 100
Light 6 "	13⅓	62 600
Heavy 6 "	16⅔	76 800
Light 7 "	20	101 000
8 "	21⅔	135 000
Heavy 8 "	26⅔	168 000
Light 9 "	23⅓	152 000
Heavy 9 "	28⅓	189 000
Extra 9 "	41⅔	268 000
Light 10½ "	35	286 000
Heavy 10½ "	45	360 000
Light 12¼ "	41⅔	377 000
Heavy 12¼ "	56⅔	511 000
Light 15 "	50	551 000
Heavy 15 "	66⅔	748 000
	CHANNELS.	
Light 6 inch.	11	41 800
Heavy 6 "	15	55 400
Light 9 "	16⅔	101 000
Heavy 9 "	23⅓	147 000
Light 12¼ "	28⅓	242 000
Heavy 12¼ "	46⅔	384 000

To find the safe load in pounds for Trenton Beams, when the beam is supported at each end, and the load is UNIFORMLY DISTRIBUTED over the span.

RULE—Divide the number given for the beam in the column headed "Co-efficient for Strength," by the distance between supports estimated in feet. The load so found will be nearly one-third of the ultimate or breaking strength of the beam. When the load is concentrated entirely at the centre of the beam, one-half the above amount must be taken.

The DEFLECTION in inches, at the middle of the span, for such distributed load, will be found by dividing the square of the span, taken in feet, by seventy (70) times the depth of the beam, taken in inches.

EXAMPLE.—What uniformly distributed load will a 12¼ inch beam of 125 lbs. per yard, and having a clear span of 15 feet, bear with safety, and what will be the deflection under this load?

Ans. $\frac{377000}{15} = 25{,}133$ lbs. = SAFE LOAD.

$\frac{15\text{x}15}{70\text{x}12\frac{1}{4}} = 0.26$ in. = DEFLECTION.

PRICES OF

ROLLED I BEAMS AND CHANNELS.

10½ inch Beams, and smaller, not over 30 feet long,........ — cents per lb.
12¼ inch Beams, not over 25 feet long .... —¼ " "
15 inch Beams, not over 20 feet long .... —½ " "

Greater lengths, ¼ cent per lb. extra for each additional 5 feet, or part of 5 feet.

Price of Channels, ¼ cent per lb. greater than that of Beams of same size.
" " Punching, ¼ cent per lb.; Plain Fitting, ¼ cent per lb.
" " Punching and Fitting on same beam, ⅜ cent per lb.
" " Wrought Fittings, — cts. per lb. Cast Separators, — cts. per lb.
" " Painting Beams or Fittings, 1/10 cent per lb.

TABLE OF STRENGTH OF RIVETED GIRDERS.

When loads or spans occur too great to admit of the use of rolled beams, it becomes necessary to employ riveted girders of greater depth. These are usually made either of **I** or box form, and the following tables enable the proper dimensions of such girders to be determined with facility.

The numbers given in the tables are the products of that portion of the SAFE DISTRIBUTED LOAD, in tons of 2,000 lbs., which can be borne by each of the component parts of the girder, viz., the stem, the angle iron, and the top and bottom flanges multiplied by the clear span in feet, and therefore the sum of the numbers given by the three tables, divided by the length of span, will give the safe load of the entire girder.

The tables are calculated assuming a maximum strain on the iron of 12,000 lbs. per square inch, and it is assumed that holes for $\frac{3}{4}''$ rivets are punched in the plates. The third table gives the strength due to each inch of *effective* width of the top and bottom plates, that is, of each inch of width in addition to the diameter of the rivet-holes.

The deflection of the girders with the distributed loads given by the tables will be, as in the case of rolled beams, equal to the square of the span in feet divided by seventy times the depth of the girder in inches.

If a less strain per square inch on the iron is desired, the load given by the tables must be proportionately reduced.

The relative strength and deflection for loads other than uniformly distributed will be determined by the same rules as in the case of rolled beams. Box girders have more stiffness sideways than those of the **I** form, and hence are used in cases where the girder is unsupported laterally. In other cases the **I** section is preferred, being more economical and more accessible for painting.

I.—PRODUCT OF LOAD IN TONS, MULTIPLIED BY SPAN IN FEET FOR EACH ⅓ INCH THICKNESS OF VERTICAL STEM PLATE.

Depth of Girder in inches.		12	13	14	15	16	17	18	19	20	21	22	23	24	25	26	27	28	29	30	31	32	33	34	35	36
Thickness of	0	9.8	11.5	13.4	15.4	17.6	20.0	22.5	25.2	28.1	31.1	34.3	37.7	41.2	44.9	48.7	52.8	57.0	61.3	65.8	70.5	75.4	80.4	85.6	90.9	96.5
	¼	8.6	10.2	12.0	13.9	16.0	18.2	20.6	23.2	26.1	28.9	31.9	35.2	38.6	42.2	45.9	49.8	53.9	58.1	62.5	67.1	71.8	76.7	81.8	87.0	92.4
top and bot-	½	7.5	9.0	10.7	12.5	14.4	16.6	18.9	21.3	24.0	26.7	29.7	32.8	36.1	39.6	43.2	46.9	50.9	55.0	59.3	63.7	68.3	73.2	78.0	83.1	88.4
	¾	6.5	7.9	9.5	11.2	13.0	15.0	17.2	19.6	22.1	24.8	27.6	30.6	33.7	37.1	40.5	44.2	48.0	52.0	56.2	60.5	65.0	69.6	74.4	79.4	84.6
tom plates,	1	5.7	6.9	8.4	9.9	11.7	13.6	15.7	17.9	20.3	22.8	25.5	28.4	31.5	34.7	38.0	41.6	45.3	49.2	53.2	57.4	61.8	66.3	71.0	75.8	80.8
	1¼	4.9	6.0	7.4	8.8	10.5	12.2	14.2	16.3	18.6	21.0	23.7	26.4	29.3	32.4	35.6	39.0	42.6	46.4	50.3	54.4	58.6	63.0	67.6	72.3	77.2
in inches.	2½	4.2	5.2	6.4	7.8	9.3	11.0	12.8	14.8	17.0	19.3	21.8	24.8	27.2	30.2	33.3	36.6	40.1	43.7	47.5	51.5	55.6	59.9	64.3	69.0	73.7

II.—PRODUCT OF LOAD IN TONS, MULTIPLIED BY SPAN IN FEET FOR FOUR CONTINUOUS ANGLE BARS, 3″ × 3″ × ½″, WEIGHING EACH 9.2 LBS. PER FOOT, CONNECTING THE FLANGES WITH THE VERTICAL WEB.

Depth of Girder in inches.		12	13	14	15	16	17	18	19	20	21	22	23	24	25	26	27	28	29	30	31	32	33	34	35	36
Thickness of	0	166	184	201	218	236	254	271	289	307	325	343	360	378	396	414	432	450	467	485	503	521	538	556	574	592
	¼	122	135	149	164	178	193	208	222	236	251	265	280	294	309	324	339	354	369	383	398	413	428	443	458	473
top and bot-	½	109	123	137	151	166	180	195	209	223	238	252	267	282	296	311	326	341	355	370	385	400	415	430	445	460
	¾	98	112	125	140	153	168	182	196	211	225	240	254	269	283	298	312	327	342	357	371	386	401	416	431	446
tom plates, in	1	87	101	114	128	142	156	170	184	199	213	228	242	256	271	285	200	314	329	344	358	373	389	402	417	432
	1¼	76	90	103	117	131	145	159	173	187	201	216	230	244	258	273	287	302	316	331	346	360	375	389	404	419
inches.	1½	68	61	94	106	120	134	147	161	175	189	204	218	232	246	261	275	289	204	319	333	348	363	377	392	406

III.—PRODUCT OF LOAD IN TONS, MULTIPLIED BY SPAN IN FEET, FOR EACH effective INCH OF WIDTH OF TOP AND BOTTOM FLANGE PLATES.

Depth of Girder in inches.	12	13	14	15	16	17	18	19	20	21	22	23	24	25	26	27	28	29	30	31	32	33	34	35	36
Thickness of top and bottom plates, in inches. ⅜	11.5	12.5	13.5	14.6	15.5	16.5	17.5	18.5	19.5	20.6	21.5	22.5	23.5	24.5	25.5	26.5	27.5	28.5	29.3	30.6	31.5	32.5	33.5	34.5	35.5
½	22.1	23.9	26.0	28.0	30.0	32.0	34.0	36.0	38.0	40.0	42.0	44.0	46.0	48.0	50.0	52.0	54.0	56.0	58.0	58.6	62.0	64.0	66.0	68.0	70.0
¾	31.7	34.7	37.7	40.7	43.6	46.4	49.9	52.6	55.0	58.6	61.6	64.5	67.0	70.6	73.6	76.6	79.6	82.6	86.6	88.6	91.6	94.6	97.6	100.6	103.5
1	40.4	44.4	48.4	52.4	56.3	60.3	64.3	68.3	72.3	76.3	80.2	84.2	88.2	92.2	96.2	100.2	104.2	108.2	112.0	116.0	120.0	124.0	128.1	132.2	136.1
1¼	48.4	53.4	58.3	63.2	68.2	73.1	78.1	83.1	93.0	93.0	98.0	103.0	107.9	112.9	117.9	122.9	127.9	132.9	137.8	142.8	147.8	152.8	157.8	162.8	167.8
1½	55.5	61.3	67.3	73.2	79.1	85.0	91.0	96.9	102.9	108.9	114.8	120.8	126.7	132.7	138.7	144.7	150.6	156.6	163.0	168.6	174.6	180.6	186.5	192.8	198.5

USE OF TABLES.—What must be the dimensions of a girder 24″ deep, having a span of 20 feet, to support a uniformly distributed load of 40 tons?

Supposing the thickness of the centre web plate to be ⅜″, and assuming for the present that the top and bottom plates will be about 1″ thick, we have from Table I., for the web 31.5 × 3 = 94.5
from Table II., for the angles 256.
                                    350.5

Product of span by load = 40 × 20 = ... 800.
Amount thus found to be sustained by web and angles........................ 350.5

Balance to be sustained by top and bottom plates 449.5
From the third table we find, under 24″ depth and 1″ thickness of flanges, the number 88; hence $\frac{449.5}{88} = 5.1$ inches of effective width of flanges, or, say, 7″ for the entire width to allow for the rivet holes punched out.

The deflection will be $\frac{20 \times 20}{24 \times 70} = .24$ inch. Of course thinner and wider top and bottom flanges may be substituted, if preferred, by using in the third table the number corresponding to the thickness desired, and in that case, if perfect accuracy is necessary, a slight change would be made in the strength furnished by the web and angles. Thus, for ¾″ flanges the strength of web would be 33.7 × 3 = 1011.1, and of the angles 269; total 370.1 instead of 350.5. The difference is unimportant.

MOULDED ROLLED WROUGHT IRON BEAMS.

The best form for wrought iron floor beams is that known as **I** beams. When the fracture of a beam of any kind is produced by vertical pressure, the fibres of the lower section are separated by extension, while at the same time those of the upper portion are destroyed by compression.

In cast iron, the resistance to compression is about 6½ to 1 of tenacity; therefore, to have the strongest section in cast iron, the bottom flange must have 6½ times the quantity of material that is contained in the top flange.

In wrought iron, although the ultimate resistance to tension is considerably greater than to compression, the amount of extension or compression, within the limits of strength which can be used in practice, is about the same for either force. Makers of rolled beams, for convenience sake, etc., have generally made the top and bottom flanges alike in weight and shape.

The strength of a rolled beam lies mainly in its vertical web, the strength being in proportion to the depth of web. The horizontal flanges give it lateral stiffness or power to resist buckling or bending side-ways.

In the Moulded Beam, the depth of web is increased, and an additional amount of material put in the bottom flange.

The entire weight of brick arches and the superimposed load comes on the lower flanges of floor beams—below the neutral line. In the case of plate girders, the entire weight is usually placed on the top—above the neutral line—as, for instance, in sustaining a brick wall.

Fairbairn lays down the rule for plate girders, that the upper flange should be larger than the lower in the ratio of 1.35 to 1; and it not only is advisable, but very convenient to have the top plate to build upon wider than the bottom.

When the weight comes on the lower flanges, as in the case of floor beams, an approximate amount of material must be

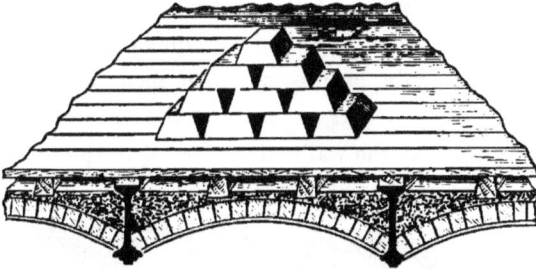

provided in the lower flange to make up for the relatively different positions of the load. On the beam, the load is at the bottom; on the girder, at the top.

For the reasons given, the Moulded Beams are stronger and more rigid than the plain beams. Rigidity prevents vibration, the avoidance of which is of great importance, for the admissible deflection of beams is limited by the amount which would cause the plastering of ceilings to crack. The limit of deflection thus allowed in beams is one-thirtieth of an inch to the foot of span—one inch in thirty feet of span.

In cast iron, a knowledge of the absolute strength or resistance to rupture of a beam is necessary. In wrought iron, a knowledge of this kind is not so important as a knowledge of the power to resist deflection. By a knowledge of the power of a beam to recover itself after the removal of a load, is ascertained what load may be placed upon it without injury to the integrity of the metal; that is, without set or permanent deflection.

The Moulded Beams contain the same width and thickness of top and bottom flanges, and the same thickness and depth of web as the standard plain beams, and have in addition an increased depth of web and an increased amount of iron underneath the usual bottom flange; the weights being in addition to the standard weights just what the moulding adds. They must necessarily give better results in points of strength and rigidity. They can, therefore, be placed farther apart than the plain beams, and thus prove more economical. Or a smaller size can be used, and be equivalent in strength to larger sizes of plain beams—a 9-inch moulded beam for a 10 inch plain beam, and so on. At the ends of the beams, where they rest on the walls, small cast iron plates are used so as to get level and solid bearings.

Strength and ornament are combined. In buildings of a public character, such as banks, offices, etc., the preferable mode of construction is to use brick arches, and leave exposed the lower flanges of the iron beams. The moulded bottom flange makes a finished appearance underneath.

The moulding adds a weight to the beams as follows:

Per lineal foot:

2¼ lbs.	3¾ lbs.	5¾ lbs.
on	on	on
4 in. Beams,	6 in. Beams,	8 in. Beams,
Light and Heavy.	Light and Heavy.	Light and Heavy.
6¼ lbs.	8 lbs.	10 lbs.
on	on	on
10¼ in. Beams,	12¼ in. Beams,	15 in. Beams,
Light and Heavy.	Light and Heavy.	Light and Heavy.

FIRE-PROOF CYLINDRICAL TILE FLOORS AND CEILINGS.

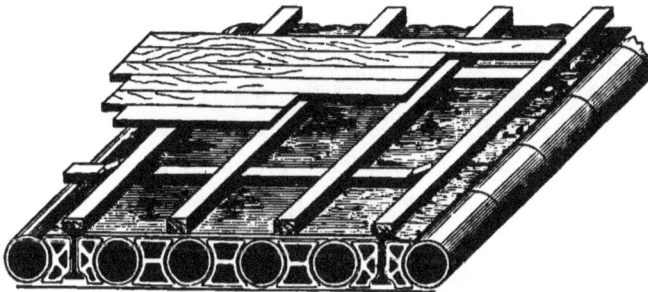

To secure a flat fire-proof ceiling, the space between the iron beams is filled in with a series of hollow cylinders, and hollow double concave binders; the abutment pieces, or end binders, being shaped to fit the lower flange of the beam. The material is of burnt clay. All parts are of equal thickness, and being cylindrical, avoids tensile strain and throws the whole weight on the compressive strength of the hollow cylinders, which form perfect arches in themselves, every inch being equal in strength to resist the pressure brought thereon. The pieces "break joints" with each other, and are laid up with thin cement. Any variation of space is provided for by a few different widths of binders.

This construction gives the strength of an arch, while, at the same time, it provides a flat, level ceiling and floor, and forms a perfect key for the plastering and hard-finish underneath. No furring or lathing is required. The wooden floor boards are laid upon wooden strips or joist; the ends of the latter fit under the top flange of the iron beams. A free circulation of air thus surrounds the joist and floor boards, and preserves them from dry-rot.

When desired or necessary, the wooden flooring may be dispensed with, and the top covered with cement, or laid with marble or stone slabs.

The forms secure the maximum of strength, durability, and compactness consistent with the greatest economy of material and labor. Mechanically the construction is correct, while the material is the only one known of such an indestructible character as to resist the attacks of fire, water, extreme changes of temperature, or all these combined. This material—the same as bricks are made of—has been proven by the experience of centuries to be the only practical fire-proof building material in existence.

Unlike the thousand and one modern conglomerations introduced under as many different names, and all equally deficient in the first elements of fire-resisting qualities, the tile used is identical with the ancient burnt clay found among the ruins of every country yet discovered.

ADVANTAGES.

A saving of WEIGHT of forty per cent., which admits of a

great reduction in weight of iron beams, thickness of side walls and foundation.

A LEVEL CEILING AND FLOOR, dispensing with furring and lathing; also with the concrete filling always necessary on the top of the old method of solid brick arches.

HEIGHT of three inches is saved between ceiling and floor.

The work can be LAID UP IN ANY SEASON OF THE YEAR.

SAVING IN TIME, as the work is dry almost as soon as laid.

STRENGTH AND ELASTICITY IN RESISTING SUDDEN IMPACT, and able to carry with safety 2,000 pounds on each superficial foot, without apparent deflection. It will, indeed, sustain a load far beyond the carrying strength of the iron beams.

No CRACKING OF CEILINGS can occur, thus avoiding the periodical patching up of plastering, and consequent repeated renewing of painting and decorations.

This floor is SOUND-PROOF, dry, and of even temperature, and will last for ages.

It is made of material abundant all over the world, and always available and cheap. It is the ONLY MATERIAL REALLY FIRE-PROOF and indestructible to time and the elements.

SPECIFICATION

OF CYLINDRICAL TILE FLOORS AND CEILINGS.

The spaces between the iron floor beams shall be filled in with hollow cylinders eight inches in diameter, and hollow double concave binders corresponding to same; the abutment pieces, or end binders, being made to fit the lower flange of the beam and to project about half an inch below the same. Any variations in spaces of beams is to be provided for by different widths of binders. The material to be red clay (or fire-clay), well burned, and made not less than three-fourths of an inch in the thinnest parts, and the outer surfaces grooved or roughened to give adhesion to the plaster. The pieces shall be about twelve inches in length each, and shall break joints with each other, and kept horizontal on the under side by being laid up on proper flat centres. The work shall be laid up with [Portland] cement mortar in proportion of one of cement to three of sand. The top of the tiles to receive a coat of same kind of cement mortar, half inch thick, laid water-tight. The under side of iron beams (plain beams) to receive a coat of gauged mortar to make them level with the tiles. Do all cutting for tie-rods; make good after gas and other pipes are laid; repair any damage done; and deliver the work ready for plastering. Furnish all scaffolding, centres, tools, materials, etc., for setting the work. Water will be provided to the contractor at a central point in the building, on all the stories, into tanks or tubs by him provided. The work to progress as the architect will direct. Make such tests of strength as may be required, by placing at the places the architect may select weights up to 1,200 lbs. on the square foot, which the tile are to support without apparent deflection.

COST OF TILE FLOORS.

Tiles, $85.00 per M., is per foot sup.			18 c.
Breakage, etc.			2
Laying up, on each 100 feet square.			
1 bbl. Portland cement	$6.00		
3 " sand	50		
		$6.50	
Labor, 1 mason, $3; 1 helper, $1.50		4.50	
Scaffolding		1.50	
Cartages		1.00	
Handling, etc		50	
			$14.00 is 14
Royalty			3
Cost per superficial foot			37 c.

Add profit 15 per cent.

TEST OF STRENGTH.

From the Engineer's Report of the test of strength of the Cylindrical Tile Floors, made at the new Capitol building at Albany, N. Y., January 21, 1874, under the direction and in the presence of the Hon. Wm. J. McAlpine, Chief Engineer:

"A section of 8 inch diameter tiles, 1 foot wide by a span of 4 feet, was built in between two wrought-iron beams placed 4 feet apart and elevated above the floor. On the centre of the section of tile a plate 12″ × 12″ was laid, and on this plate a total weight of 3,604 lbs. was placed. This weight remained on over night—say 14 hours. No perceptible deflection took place; and the section after the removal of the weight was apparently in as good condition as before the test was made.

"The weight of 3,604 lbs. in the centre is equivalent to about double the weight equally distributed, or 7,200 lbs., which would give 1,800 lbs. per foot superficial. Say only one-half of this result be taken, 900 lbs., so as to be absolutely on the safe side. Now as a variable load equal to a crowd of people may be taken at 120 lbs. per foot floor surface, there would be a safe margin of over seven times the required strength.

"For warehouses the load to be carried should be computed at 350 lbs. and upwards. The test of the tiles show an ample margin.

"The test shows that for all practicable purposes the Cylindrical Hollow Tile Floors (constructed as shown in the illustration) have ample strength for the purposes for which they are intended, and far beyond the carrying strength of the iron beams of the usual size and placed the usual distances apart.

"A strength beyond this is useless. It is asserted that the section already tested will bear a weight applied in the centre of 8,000 lbs. and upwards as its absolute strength. This would be an enormous load. But beyond the results already attained nothing further can be desired in strength.

(Signed)   PETER HOGAN, C.E."

IRON ROOFS.

As many elaborate treatises on Roofs are published, more than a general reference will not be made. Iron trusses for rafters combine lightness, strength, durability, and consequent economy. For simplicity and economic arrangement of material, Figs. 1 and 2 are most generally adopted in practice.

For principals, **I** beams make very good rafters, and in light trusses, **T** bars, with or without a plate rivetted to the upper flange, answer every purpose. Struts may be made of **T** bars or angle iron, as these forms afford great facility for attachment to the rafters.

Fig. 3 shows the modification of the ordinary King and Queen Truss as adapted to wrought iron. Figs. 4 and 5 are circular roofs, the details being very similar to those for double pitch roofs.

9

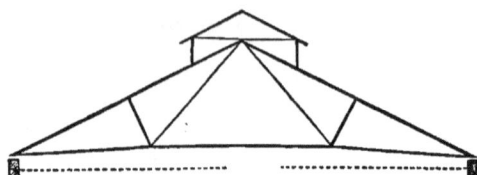

Say 57 ft.

FIG. 1.

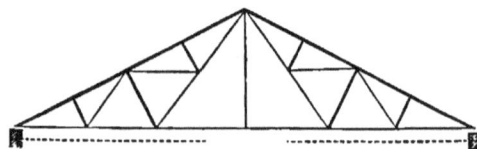

Say 112 ft.

FIG. 2.

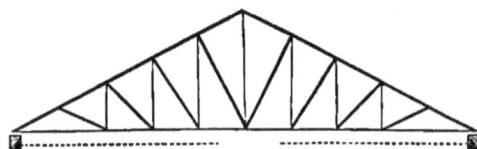

Say 134 ft.

FIG. 3.

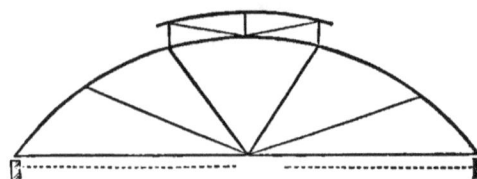

Say 63 ft.

FIG. 4.

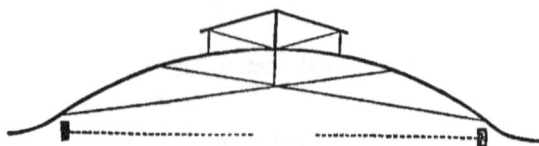

Say 45 ft.

FIG. 5.

Ties may be of flat or round bars, attached by eyes and pins or screw ends. Care should be especially taken to properly proportion the dimensions of eyes and pins to the strains upon them. A very good and safe rule in practice is to make the diameter of the pin from three-fourths to four-fifths of the width of the bar in flats, and one and one-fourth times the diameter of the bar in rounds, giving the eye a sectional area of fifty per cent. in excess of that of the bar. The thickness of flat bars should be at least one-fourth of the width, in order to secure good bearing surface on the pin, and the metal at the eyes should be as thick as the bars on which they are upset.

The details of roof trusses vary to suit the character of the work, and the sections of iron employed. The heel of the rafter usually rests on the wall, in a cast-iron skewback fitted to the beam, and sloping to the angle required by the pitch of the roof (see Fig. 6). The struts are attached to the rafters by cast caps, or by wrought strap plates (see Fig. 7), and the joint at their feet made either for pin or screw connections (see Fig. 8). The peak is joined by wrought plates and bolts, the beams having been cut to the required angle (see Fig. 9).

FIG. 6.—HEEL.

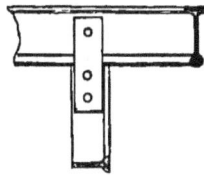

FIG. 7.—STRUT HEAD.

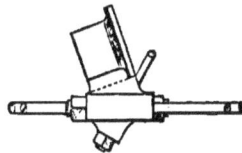

FIG. 8—STRUT FEET.

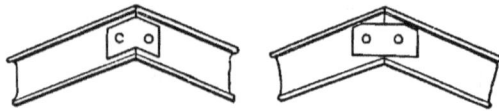

FIG. 9.—PEAKS.

In roofs of wide span, provision for expansion of the iron, due to changes of temperature, must be made by resting the skewback of one end of the truss on a cast wall-plate, with rollers interposed to permit of the sliding of the heel without straining the wall; but this precaution is not necessary in roofs of sixty feet span or less. Careful experiments have proved that an iron rod one hundred feet long will vary about $\frac{1}{10}$ of a foot for a change of temperature of 150 degrees Fahren., and as this is the greatest range to which iron beams and rods in a building would probably be subjected in this climate, compensation to that amount would be sufficient for all purposes. For sixty feet span the vibration of each wall would then be only $\frac{15}{1000}$ of a foot either way from the perpendicular, a variation so small and so gradually attained that there is no danger in imposing it upon the side walls by firmly fastening to them each heel of the rafter. Expansion is also provided against by fastening down one heel with wall bolts, and allowing the other to slide to and fro on the wall-plate, without rollers, as shown in Fig. 10.

FIG. 10.

In estimating the strains on roofs, the weight of the structure itself, as well as the loads to be supported, must be taken into account. Tredgold's assumption of the total maximum vertical load at forty pounds per square foot of horizontal surface is usually considered sufficiently high; but, if a floor or ceiling is suspended to the tie beam, or should the under side of the rafters be boarded and plastered, it is evident that these additional weights require more strength in the roof for their support.

For ordinary roofs of short span, thirty pounds per square foot is quite enough, however; and for long spans over sixty feet, thirty-five pounds will be sufficient to provide for with the factors of safety in the material that are usually adopted. The stress upon each member of the truss having been found by any of the methods of calculation preferred, the sectional areas may be found by taking the tensile strength of good wrought iron at 10,000 pounds per square inch, and the compressive resistance of beam or shape iron at from 6,000 or 8,000 for the same unit of section. The smaller or counter-balance rods ought to be made strong enough to resist strains induced by wind pressure on one side of the roof only—the other half being unloaded.

Lateral braces, as in Fig. 11, should be provided in each end panel of straight roofs, as well to secure the roof during erection as to provide an abutment that will uphold the whole in case of fire or accident. From the panels so braced, tie rods run to each of the other rafters, and, with the purloins, unite the roof into a firm and compact whole. The gable walls are sometimes used to anchor the end rods into, but the method shown in the figure is that which is generally preferred.

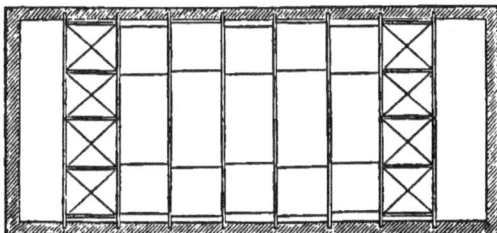

Fig. 11.

Main rafters may be spaced from four to twenty feet apart, the spacing being regulated by the size of the purlin, and this again by the material used for covering. For slate on iron purlins a convenient spacing is about eight feet between centres of rafters, the angle iron purlins being put at seven to fourteen inches apart, according to the size of the slate used, and notched at the ends into the flanges of the rafters. They are held in place by tie rods that reach from rafter to rafter the entire length of the building, three or four rows of these rods being placed between peak and heel, at from six to eight feet intervals. On the iron purlins the slate is laid directly and held down by copper or lead nails, clinched around the angle bar, as shown in Fig. 12.

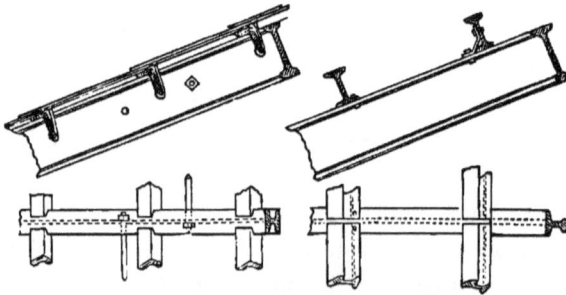

FIG. 12.—PURLINS.

When greater intervals are used in spacing rafters, the purlins may be light beams fastened on top or against the sides of the principals with brackets,—allowance always being made for longitudinal expansion of the iron by changes of temperature. On these purlins are fastened wooden jack rafters, carrying the sheathing boards or laths, on which the metallic or slate covering is laid, in the usual manner; or sheets of corrugated iron may be fastened from purlin to purlin, and the whole roof be entirely composed of iron.

When the rafters are spaced at such intervals as to cause too much deflection in the purlines, they may be supported by a light beam, placed midway between the rafters and trussed transversely with posts and rods. These rods pass through the rafters and have beveled washers, screws, and nuts at each end for adjustment (see Fig. 13).

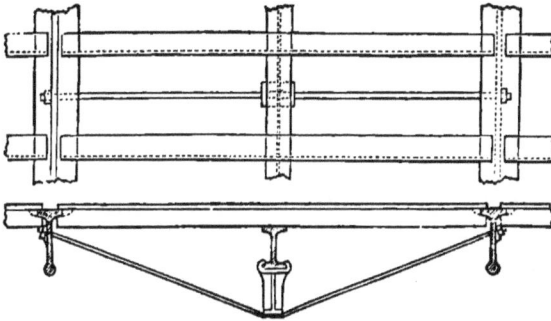

FIG. 13.—IRON PURLINS TRUSSED.

By alternating the trusses on either side of the rafter and slightly increasing the length of the purlins above them, leaving all others with a little play in the notches, sufficient provision will be made for any alteration of length in the roof, due to changes of temperature.

When wooden purlins are employed, they may be put between the rafters, and held in place by tie rods, or on top, and fastened to the rafter by brackets (see Fig. 14).

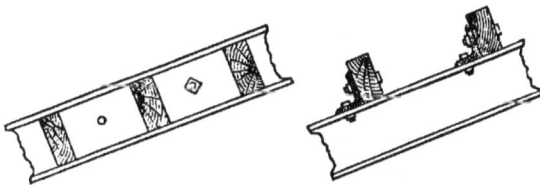

FIG. 14.—WOODEN PURLINS.

The sheathing boards and covering are then nailed down on top of all in the usual manner.

In Fig. 15 the purlins of angle iron carry wooden strips, to which are nailed the sheathing boards and covering material.

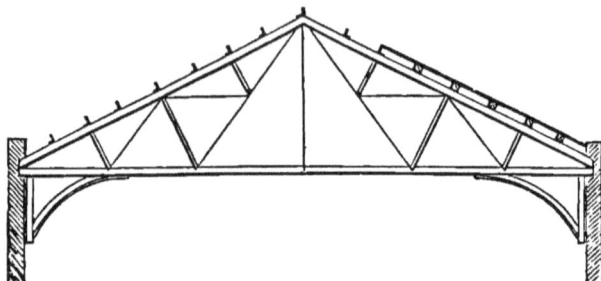

Fig. 15.

Light arches of tiles or hollow bricks may be turned on the lower flanges of smaller transverse beams, as described for floors.

When desired ventilators or lanterns are added along the ridge of the roof, the attachments being securely made to the rafters by wrought brackets and bolts, and the bracing effected in a cheap and thorough manner by two tie rods that run from the peak of the rafter to the angle between the post and rafter of the ventilator, the covering material being attached as described for the main rafters.—*Phœnix Iron Works.*

MANSARD ROOFS.

The best fire-proof Mansard roof, and one that has been extensively used, is constructed with a continuous bed-plate, top beam, and uprights, all bolted together, thus forming a rigid framework of iron and then filled in with hollow baked clay tiles.

The front of the building is carried up vertically to the raking line of the Mansard. The preferable way, even when the front is of cast iron, is to build up a twelve-inch brick wall. On this is laid a cast-iron plate, channel-shaped, 10 inches wide by $\frac{1}{2}$ inch thick, the edges turning up 2 inches. The top plate, same shape, 6 inches wide and $\frac{1}{2}$ inch thick. The rafters are **H** shape, $3 \times 2 + \frac{1}{2}$, with an open additional web projecting

3 inches at top and 7 inches at bottom. In this web wrought iron pins are cast—rivets, with the heads in the casting—at the proper distances, as required by the size of slate to be used. The rafters are bolted to the bottom plate, each with two $\frac{1}{2}$ inch bolts and nuts, and to the top plate with one $\frac{3}{4}$ inch bolt and nut. The rafters should not be placed more than 4 feet apart and set up on an inclined line. Purlins of $1 \times \frac{3}{16}$, are placed on the outside of the rafters. The purlins are first laid on the rafters and marked for punching. The purlins, after being punched with holes to correspond with the positions of the pins, are laid on, and the pins partly sniped off and hammered down.

All necessary straps, anchors, etc., must be furnished.

The iron frame to be filled in with hollow baked clay-tiles, $4\frac{3}{4}$ inches wide, laid up in cement. The back of the tiles —facing the room—to be plastered and hard finished.

The slate on the iron purlin bars to be hung with suitable copper wire, carefully twisted, two wires to each slate, and the slate made to lay flat.

In case of a required alteration from a wooden mansard to a fire-proof construction, all the present wood-work must be taken down and removed; a board partition put up about ten feet back from the front; and such scaffolding, etc., as necessary, provided; all cutting and bracketing done that may be necessary; the floor, etc., pieced out, to correspond with the present flooring, etc.; do all patching and piecing out and making good that may be required; the iron and tile and plastering work, as in the case of a new roof; take off the present galvanized iron-top cornice, etc.; carefully remove the slate; re-hang the old slate with copper wire; put up galvanized dormer windows in place of the present wooden ones, to be exactly alike in design as the ones now on the building, the dormers to be braced with iron and the top of the frame covered with corrugated galvanized iron No. 20; put galvanized iron cheeks and window heads to the window frames inside; all bracing to be done in iron; make the tin work on roof good where injured during the progress of the work; replace the iron top cornice; do generally all the galvanized iron work and tin work requisite to be done in and about the work, and leave all water-tight at completion; do all carpentering and mason work requisite to be done; cover in the opening each night with canvas of sufficient size as to protect the building and goods stored therein from injury during the progress of the work; all the exposed iron work to be painted two coats best white lead and linseed oil paint; outside the color to correspond with the present color of front; inside work painted white or such color as directed.

PIG IRON.

Highest and lowest quotations of Pig Iron, at New York, per ton, from 1825 to 1875 inclusive:

Year.	Lowest.	Highest.	Year.	Lowest.	Highest.
1825............	$35 00	$75 00	1851.................	$19 00	$25 00
1826............	50 00	70 00	1852.................	19 00	31 00
1827	50 00	55 00	1853.................	28 50	38 00
1828................	50 00	55 00	1854.................	32 00	41 50
1829................	40 00	50 00	1855.................	27 50	36 00
1830................	40 00	50 00	1856.................	29 00	37 00
1831................	40 00	47 50	*1857.................	29 00	36 00
1832................ .	40 00	47 50	1858....	23 00	27 00
1833................	37 50	47 50	1859.................	23 00	30 00
1834................	38 00	45 00	1860.................	22 50	27 00
1835................	38 00	42 50	1861......	20 00	25 00
1836	40 00	60 00	*1862.................	21 00	33 50
*1837................	40 00	70 00	*1863....	30 00	45 00
*1838................	37 50	50 00	*1864.................	47 50	70 00
1839................	37 50	42 50	1865.................	35 00	55 00
1840................	34 00	38 00	1866.................	42 00	55 00
1841................	32 00	37 50	1867.................	38 00	49 00
1842................	23 50	35 00	1868.................	35 00	45 00
1843..	22 50	32 00	1869.................	39 00	42 00
1844......	30 00	35 00	1870.................	31 00	35 00
1845................	30 00	42 00	1871.................	30 00	36 00
1846................	35 00	42 50	1872.................	36 00	53 00
1847................	33 00	42 50	1873.................	33 00	48 00
1848................	26 50	37 50	1874......	26 00	33 00
1849......	22 50	26 00	1875........	24 00	27 00
1850................	22 00	24 00			

No. 1 American.

* Years of bank suspension.

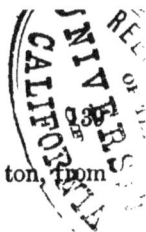

Price of No. 1 Anthracite Foundry Pig Iron, per ton, from 1844 to 1875 inclusive:

Year.	Av.	Year.	Av.
1844	$26	1860	$23
1845	29	1861	20
1846	28	1862	24
1847	30	1863	35
1848	27	1864	59
1849	23	1865	46
1850	21	1866	47
1851	21	1867	44
1852	23	1868	39
1853	36	1869	41
1854	37	1870	33
1855	38	1871	35
1856	27	1872	49
1857	26	1873	47
1858	22	1874	30
1859	23	1875	26

BAR IRON.

Highest and lowest quotations of Bar Iron, at New York, per ton, from 1825 to 1875 inclusive:

Year.	Lowest.	Highest.	Year.	Lowest.	Highest.	
1825	$85 00	$115 00	1851	$33 00	$40 00	
1826	85 00	100 00	1852	34 00	55 00	
1827	77 00	90 00	1853	50 00	75 00	
1828	77 50	82 50	1854	62 50	77 50	
1829	75 00	82 50	1855	65 00	65 00	
1830	72 50	77 50	1856	50 00	65 00	
1831	72 00	80 00	*1857	53 00	62 00	
1832	72 00	75 00	1858	44 00	55 00	
1833	72 00	75 00	1859	42 50	48 00	
1834	71 00	75 00	1860	41 00	43 00	
1835	67 00	75 00	1861	40 00	50 00	
1836	67 00	72 50	*1862	50 00	70 00	
*1837	75 00	100 00	*1863	72 50	75 00	
*1838	85 00	105 00	*1864	105 00	220 00	
1839	85 00	90 00	1865	95 00	180 00	
1840	82 50	95 00	1866	94 00	110 00	
1841	70 00	82 50	1867	85 00	105 00	
1842	62 50	75 00	1868	90 00	92 50	
1843	65 00	62 50	1869	85 00	90 00	
1844	55 00	60 00	1870	77 50	95 00	From Store.
1845	57 50	62 50	1871	75 00	90 00	
1846	62 50	80 00	1872	87 50	116 00	
1847	75 00	80 00	1873	68 00	100 00	
1848	50 00	70 00	1874	56 00	68 00	
1849	42 50	50 00	1875	56 00	60 00	
1850	40 00	45 00				

* Years of bank suspension.

COAL.

The following table gives the wholesale price of Anthracite Lump Coal, at the port of New York, at the opening of navigation, from 1852 to 1875 inclusive:

Year.	Price per ton.	Year.	Price per ton.
1852	$3 80	1864	$8 00
1853	3 95	1865	10 30
1854	4 50	1866	7 10
1855	5 00	1867	5 60
1856	4 50	1868	4 15
1857	4 00	1869	4 40
1858	3 50	1870	4 65
1859	3 35	1871	6 00
1860	3 60	1872	4 10
1861	3 50	1873	4 75
1862	3 20	1874	5 00
1863	4 40	1875	5 10

GOLD.

The following table gives the highest and lowest prices of Gold, from 1862 to 1875 inclusive:

Date.	Lowest.	Highest.	Date.	Lowest.	Highest.
1862	par to	137	1869	119¼	144¾
1863	122¼	172½	1870	110¼	122¼
1864	151½	285	1871	108¼	115⅝
1865	128⅝	234½	1872	108¼	115⅝
1866	125	167¾	1873	106½	119½
1867	132	146⅜	1874	109	114⅜
1868	133	150	1875	111¾	117¾

A STOCK COMPANY.

In manufacturing enterprises it will be found easier to obtain the necessary capital from a number of persons by subscriptions in stock, than to get it from an individual in a general or special copartnership.

The following is given as a complete form of organization for such company, under the laws of the State of New York:

SUBSCRIPTION LIST OF THE ———— IRON WORKS.

The parties subscribing hereto, being desirous of taking shares of stock in a company to be organized by the above name, under the general manufacturing law of the State of New York, passed February 17, 1848, and the Acts amendatory thereof, do hereby mutually agree, each with the other, as follows:

I. That we will take the number of shares of stock in the said ———— Iron Works set opposite our names respectively.

II. We hereby agree to pay for the said shares of stock subscribed for by us respectively, as follows, viz.: Twenty-five per cent. thereof on the day after the organization of said company, by the filing of the Certification of Incorporation as provided for by law, and the remaining seventy-five per cent. thereof as the same may be called for by the Board of Trustees of said company, in sums not exceeding twenty-five per cent. upon each call.

Dated, ————, 187

Names of Subscribers.	Residence.	No. of Shares.

CERTIFICATE OF ORGANIZATION.

Charter.

The undersigned have this day formed a corporation, under and in conformity with a statute of the State of New York, entitled an Act to authorize the Formation of Corporations for Manufacturing, Mining, Mechanical or Chemical Purposes, passed February 17th, 1848, and the acts amendatory thereof; and in compliance with the requisitions of the aforesaid Acts, we do hereby certify as follows:

First.—The corporate name of the said company is the "——— Iron Works."

Second.—The object for which said company is formed is the manufacture and sale of iron work for building purposes, and to do a general iron-founding and machinery business, and the manufacturing of articles incidental thereto.

Third.—The amount of capital stock of said company is one hundred and fifty thousand dollars.

Fourth.—The number of shares of which said stock shall consist is fifteen hundred, of one hundred dollars each.

Fifth. The number of trustees shall be six; and the names of the trustees who shall manage the concerns of the company for the first year are —————————————————————, all residing in ———.

Sixth.—The town and county in which the operations of said company are to be carried on is ———.

Seventh.—The term of existence of said company is to be fifty years.

Dated, ———, 187

(Signed)   × L. S.
                                         ×   "
                                         ×   "
                                         ×   "
                                         ×   "
                                         ×   "

State of New York, ⎱
——— County of——— ⎰ ss:

On this ——— day of ———, 187 , before me came ——— ———————————————————, to me personally known to be the individuals described in, and who executed, the foregoing Certificate of Incorporation; and they severally acknowledged, each for himself, that they executed the same for the purposes therein set forth.

(Signed)   —————————,

[L. S.]   Notary Public.

Endorsed: Filed, ———, 187 .

State of New York, ⎫
Office of the Secretary of State. ⎬

This is to certify that the Certificate of Incorporation of the
" ———— Iron Works," with acknowledgment thereto annexed,
was received and filed in this office on the —— day of ————,
187 .
Witness my hand and seal of office of the Secretary of State,
at the city of Albany, this ———— day of ————, one
thousand eight hundred and seventy ————.
[L. S.]   (Signed)   ———————————,
                                        Secretary of State.

State of New York, ⎫
———— County of ———— ⎬

This is to certify that the Certificate of Incorporation of the,
" ————Iron Works," with acknowledgment thereto annexed,
was received and filed in this office on the —— day of ————,
187 .
Witness my hand and seal of office of County Clerk, at ————
this —— day of ————, one thousand eight hundred and
seventy ——.
[L. S.]   (Signed)   ———————————,
                                        County Clerk.

First Meeting of Trustees.

At a meeting of the Trustees of the " ———— Iron Works,"
held at the office of the company, on the —— day of ————,
187 , at 12 m., present —————— Mr. ———— was appointed
Chairman and Mr.———— Secretary.

The Certificate of Organization was read and approved.

On motion of ————, duly seconded, it was resolved to
proceed to the election of officers.

The Chairman appointed ———— and ———— as tellers, who
received the ballots and reported that there were six votes
cast for ———— as President ; the same number for ———— as

Vice-President; the same number for —— as Treasurer; the same number for —— as Secretary; and the same number for —— as Manager; all of whom were thereupon declared to be unanimously elected to fill the designated offices for one year, and until others should be elected in their stead.

Here, Mr. —— assumed the Presidency, and Mr. —— the Secretaryship.

On motion of ——, duly seconded, it was resolved that the Chair appoint three Trustees to draft By-Laws, whereupon he appointed Messrs. —— and ——, who presented the following, which were read and unanimously adopted:

[See By-Laws.]

On motion of ——, duly seconded, it was resolved that the subscriptions to the capital stock be called in, payable to the Treasurer, in four instalments of 25 per cent. each; the first on the ——th inst.; the second on the ——th day of —— next; the third on the ——th day of —— next; and the fourth on the ——th day of —— next. If any party desired to pay their subscription in full, the Treasurer was authorized to allow interest at and after the rate of 7 per cent. per annum for all sums paid in advance.

On motion of ——, duly seconded, the —— Bank was selected as the depository of the funds of the Company.

On motion of ——, duly seconded, the salary of the Manager was made $—— per annum, payable monthly.

On motion of ——, duly seconded, the following Trustees were appointed an executive committee, viz.: —— and ——.

On motion of ——, duly seconded, it was resolved that this Company now proceed vigorously in perfecting the arrangements for business; and also proceed with such expenditures for buildings, tools, materials, etc., as in the opinion of the Trustees may be warranted in view of the funds to be received and the prospective state of trade.

It was also resolved that 250 copies of the proceedings of this meeting be printed, together with the By-Laws, Charter, etc., in pamphlet form; and the Secretary directed to furnish each subscriber of stock with one copy.

Adjourned to meet on the —— day of —— at —— p. m.

(Signed) —————————————,

Secretary.

By-Laws of the —— Iron Works.

Article I.

MEETINGS OF STOCKHOLDERS.

1. All meetings of stockholders shall be held at the office of the Company, in the ——, and the annual meeting for the election of Trustees shall be held the first Monday in February, at 12 o'clock at noon, and the polls shall be kept open one hour. If for any cause an election of Trustees shall not be had on the day above designated, it may be held on any subsequent day, to be fixed by the Board of Trustees.

2. Notice of all meetings of Stockholders shall be given at least ten days prior to such meeting, by advertising the same in at least —— newspaper published in ——, and notices thereof sent to each Stockholder to his residence or address, as it appears on the books of the company.

3. All elections by the Stockholders shall be by ballot; Stockholders may vote in person or by a written proxy, and each Stockholder shall be entitled to as many votes as he represents shares of stock; and the persons receiving the greatest number of votes shall be Trustees for one year, and until their successors shall have been elected.

4. Special meetings of the Stockholders may be called by the President or any two of the Trustees, when deemed necessary, of which five days' notice shall be given to each Stockholder in the manner provided by section 2.

10

Article II.
THE BOARD OF TRUSTEES.

1. The Board of Trustees shall consist of —— members, a majority of whom shall constitute a quorum for the transaction of business.

2. All meetings of the Board of Trustees shall be held at the office of the Company, in ————.

3. In case of failure to hold any election, the Trustees shall hold over and continue in office with full authority until a new election is held.

4. No person shall be a Trustee who is not the holder or owner of at least ten shares in the capital stock of this Company.

5. No Trustee, as such, shall receive any salary or compensation for his services; but this is not to preclude any Trustee from holding any other office in the said Company, or performing any services for said Company, and receiving compensation therefor.

6. Stated meetings shall be held on the first Monday in each month, and special meetings may be held upon the call of the President, or any two Trustees, due notice thereof being given by the Secretary to all the members, either in person or by mail.

7. The order of business of the meetings of the Board of Trustees shall be conducted according to usage.

8. The officers of the Company shall consist of a President, Vice-President, Secretary, Treasurer, and Manager, and any two of these offices may be combined in one person.

9. The Board of Trustees, as soon as may be after their election, shall hold a meeting and elect by ballot or otherwise a President, Vice-President, Secretary, Treasurer and Manager, who shall hold their offices for the ensuing year, and until their successors shall have been elected and duly qualified to enter upon their respective duties; they shall also appoint an

Executive Committee, to consist of two Trustees with the President.

10. The Board of Trustees shall fix the compensation of the officers; they shall declare such dividends from the net earnings or profits of the Company when, and as often as, the state of the funds will warrant; they shall, for cause, remove any officer of the Company, but no officer shall be removed until after investigation and a concurrence of a majority of the Board of Trustees.

11. They shall select a bank or depositaries, in which all the moneys of the Company shall be deposited, to the credit of the ———— Company, subject to the draft of the Company, signed by the President and Treasurer, or the Vice-President and Treasurer, and made payable to the order of the party or parties to whom it is to be paid, when practicable.

12. They shall make a report and render an account to the Stockholders at their annual meeting, showing in detail the situation of the property and financial affairs of the Company.

13. They shall have power to fill any vacancies which may occur by death, resignation, or otherwise (in the interval between the annual meetings of Stockholders), in the Board of Trustees and Executive Committee, and in the offices of President, Vice-President, Secretary, Treasurer and Manager.

14. They shall appoint three Inspectors of Election to receive the ballots from Stockholders for Trustees, prior to their annual meeting.

Article III.

EXECUTIVE COMMITTEE.

The Executive Committee shall superintend the finances of the Company, examine and audit the accounts; they shall have power to make temporary loans of surplus funds, and attend to such duties as may be necessary during the recess of the Board of Trustees, or may be designated to it by them;

they shall keep minutes of all their proceedings, and report the same to the Board of Trustees.

Article IV.

PRESIDENT.

1. It shall be the duty of the President to preside at all meetings of Stockholders and Trustees (except those convened to remove him or inquire into his official conduct), to sign all documents and contracts authorized by the Board of Trustees; to sign all checks, notes and certificates of stocks, and to perform all such duties usually incidental to such office and required by the provisions of the Act of Incorporation and these By-Laws.

2. In case of sickness or absence of the Secretary, Treasurer or Manager, he shall appoint some person to perform the duties of either until the Board of Trustees shall be convened.

Article V.

VICE-PRESIDENT [AND MANAGER].

1. It shall be the duty of the Vice-President to attend to the business of the Company (Sundays and holidays excepted); to attend to the estimating and procuring of work, and to the execution of the same; to the employing of labor and the proper mechanical conduction of the iron works; to the purchasing of materials for the business, and shall generally exercise a supervision and control over the affairs of the Company, subject to the approval of the President and directions of the Board of Trustees. In the absence of the President, he shall preside at all meetings of Stockholders and Trustees.

Article VI.

SECRETARY.

1. It shall be the duty of the Secretary to be in attendance at the office of the Company during business hours; to give the necessary notice of all meetings of Stockholders and Board

of Trustees; he shall record the proceedings of the same in a book to be kept for that purpose; shall keep all proper books of accounts for the business of the Company, with a Stock Ledger, Transfer Book, and such other books or papers as the Trustees may direct; register and sign (with the President, and countersigned by the Treasurer) all certificates of stock, and generally shall perform such services and duties as usually appertain to his office in a corporate body, and are required by the provisions of the Act of Incorporation; all the books, papers and correspondence shall be kept in the office of the Company, and considered in his possession and charge, but open at all reasonable times during business hours to the inspection of Trustees.

Article VII.

TREASURER.

It shall be the duty of the Treasurer to attend to all collections, receive and deposit all moneys where directed, and to pay and dispose of the same under the direction of the Board of Trustees; sign all checks, drafts and notes; sign all certificates of stock with the President; keep correct accounts of the same, and give his time and attention to the duties of his office. He shall keep his bank account in the name of the Company, and shall render a statement of his cash account at each regular meeting of the Board of Trustees. He shall at all times exhibit his books and accounts and papers to any Trustee upon application at the office during business hours.

Article VIII.

MANAGER.

It shall be the duty of the Manager to attend daily to the construction in a proper and right manner of all work; to see and know that every part thereof is made of the proper material, in the right manner, and of good workmanship; to make estimates, receive work, employ labor, and superintend the mechanical departments of the Company.

Article IX.

CERTIFICATES OF STOCK.

1. The Certificates of Stock shall be numbered and registered as they are issued; they shall exhibit the holders' name and number of shares, and shall be signed by the President and Secretary, and countersigned by the Treasurer, and have the seal of the Company affixed thereto.

2. Each Certificate of Stock shall express upon its face that the share or shares thereby represented are full-paid stock, and not liable to further calls or assessments.

3. The said Certificates shall be in the usual form.

4. Transfers of Stock shall be made on the books of the Company in the presence of the President or Secretary, or authorized officer or agent, upon the surrender of· the Certificate, either by the holder in person or by attorney, and the surrendered Certificate shall be cancelled and pasted on the margin in the book from whence it was taken when issued.

5. The Transfer Book shall be closed at least three days previous to an election, or the payment of dividends, and the dividend shall be paid to the Stockholders standing on record at the closing of the books.

6. If any person claim a Certificate of the Stock of this Company in lieu of one lost or destroyed, he shall make an affidavit of the fact, and state the circumstances of the loss or destruction, and he shall advertise in one or more of the daily newspapers, to be designated by the President, for the space of one week, an account of the loss or destruction, describing the Certificate, and calling upon all persons to show cause why a new Certificate shall not be issued in lieu of that lost or destroyed; and he shall transmit to the Company his affidavit and the advertisement above mentioned, with proof of its due

publication, and shall give to the Company a satisfactory bond of indemnity against any damage that may arise from issuing a new Certificate; whereupon the President shall issue a new Certificate of the same tenor and amount with that said to be lost or destroyed, and specifying that it is in lieu thereof.

Article X.

SEAL.

A suitable seal, having the words "——— Iron Works ———," with such other device as the Trustees shall select, shall be provided, which shall be under the charge of the President, and the affixing of the seal to contracts and instruments, together with the signatures of the President and Treasurer, shall bind the Company.

The affixing of the Seal, however, to contracts for iron work, etc., to be executed, such as are usually drawn up by architects, engineers, etc., shall not be necessary; the signatures of the President (or Vice-President) and Treasurer will alone be required. In signing any contract for work amounting to under $50,000, the signature of either the President, Treasurer or Manager shall be sufficient and binding.

Article XI.

BY-LAWS.

These By-Laws shall not be altered, except by the consent of two-thirds of the whole Board of Trustees; and all proposed amendments or alterations shall be submitted to the Board, in writing, at a previous meeting to that at which the action of the Board shall be had thereon, and previous notice in writing shall be given by the Secretary to each Trustee of the Company of the contemplated amendments, and the time when they will be passed upon.

———, 187 .

OPINION.

I have considered the papers submitted to me relating to the organization of the —————— Iron Works, and am of opinion that the certificate of incorporation of said Company is drawn, executed and filed in conformity with the requirements of the General Manufacturing Corporation Acts of February 17, 1848, and of the acts amendatory thereof, and that said Company is duly organized under said acts, and entitled to all the powers and privileges accorded thereby.

I have examined the minutes of the organization of the Company, and the preparatory subscription agreement for forming the Company, and they seem to be sufficient in form and according to law.

This Company being thus duly incorporated, the stockholders are under no personal liability except as the acts in question provide, viz.:

The Stockholders are severally liable for all the debts of the Company (each to the amount of his stock) until the capital is all paid in and a certificate thereof duly made and recorded.

The Stockholders are always jointly and severally liable for all debts due to laborers, servants and apprentices for services performed for the corporation.

There are liabilities in addition on the *Trustees;* they cannot make loans to Stockholders, nor make false statements in any public report or notice, nor allow indebtedness beyond the amount of the capital stock, nor declare a dividend reducing its capital; and they must not omit to file and publish the annual statements of the condition of its affairs, as required by the statutes. These, however, are plain prohibitions, applicable to *Trustees* only, and not embraced in the ordinary liability of mere Stockholders.

Attorney, etc.

————————— 187 .

NOTE.—Small and cheap editions of the act for the formation of companies are published, giving in epitome their privileges and restrictions, and arranged with a special view to convenience and conciseness.

LIMITED LIABILITY COMPANIES.

" An Act to provide for the Organization and Regulation of certain Business Corporations, passed by the Legislature of New York, June 21, 1875," provides for two classes of corporations, to be known respectively as:

1. Full liability companies.
2. Limited liability companies.

In "full liability companies" all the stockholders are severally and individually liable to the creditors of the company in which they are stockholders for all the debts and liabilities of such company.

In "limited liability companies," the name of the company must, in every case, have as its last word the word "limited." All the stockholders are severally and individually liable to the creditors of the company in which they are stockholders, to an amount equal to the amount of stock held by them, respectively, for all debts and contracts made by such company, until the whole amount of capital stock fixed and limited by such company has been paid in.

It repeals none of the general acts for the formation of corporations theretofore passed.

A manufacturing company has the choice of organizing under the Act of February 17, 1848, or under the Act of June 21, 1875. A reading and comparison of the two laws will enable any one to understand their respective peculiarities, and to decide under which to organize.

EXTRACTS FROM

THE BUILDING LAW,

(Passed April 20, 1871),

OF THE CITY OF NEW YORK, RELATING TO IRON WORK.

.on or
wooden
girders,
and bearing
weight of
same.

§ 7.................................In case iron or wooden girders supported upon iron or wooden columns are substituted in place of partition walls, the building may be fifty feet in width, but not more; and if there should be substituted iron or wooden girders supported upon iron or wooden columns, in place of the partition walls, they shall be made of sufficient strength to bear safely the weight of two hundred and fifty pounds for every square foot of the floor or floors that rest upon them, exclusive of the weight of material employed in their construction, and shall have a footing course and foundation wall not less than sixteen inches in thickness, with inverted arches under and between the columns, or two footing courses of large well-shaped stone, laid crosswise, edge to edge, and at least ten inches thick in each course, the lower footing course to be not less than two feet greater in area than the size of the column; and under every column, as above set forth, a cap of cut granite, at least twelve inches thick and of a diameter twelve inches greater each way than that of the column, must be laid solid and level to receive the column..................................

Isolated
piers,
how con-
structed.

§ 10. Every isolated pier less than ten superficial feet at the base, and all piers supporting a wall built of rubble stone or brick, or under any iron beam or arch girder, or arch on which a wall rests, or lintel supporting a wall, shall, at intervals of not less than thirty inches in height, have built into it a bond stone not less than four inches thick, of a diameter each way equal to the diameter of the pier, except that in piers on the street front, above the curb, the bond stone may be four inches less than the pier in diameter ;.... and the walls and piers under all compound, cast iron or wooden girders, iron or other columns, shall have a bond stone at least four inches in thickness, and if in a wall at least two feet in length, running through the wall, and if in a pier, the full size of the thickness thereof, every thirty inches in height from the bottom, whether said pier is in the wall or not, and shall have a cap stone of cut granite, at least twelve inches in thickness, by the whole size of the pier, if in a pier, and if in a wall, it shall be at least two feet in length, by the thickness of the wall, and at least twelve inches in thickness. In any case where any iron or other column rests on any wall or pier built entirely of stone or brick, the said column shall be set on a base stone of cut granite, not less than eight inches in thickness by the full size of the bearing of the pier, if on a pier, and if on a wall, the full thickness of the wall................

§ 12. In no case shall the side, end, or party wall of any building be carried up more than two stories in advance of the front and rear walls. The front, rear, side, end and party walls of any building hereafter to be erected shall be anchored to each other every six feet in their height by tie anchors, made of one and a quarter inch by three-eighths of an inch wrought iron. The said anchors shall be built into the side or party walls not less than sixteen inches, and into the front and rear walls at least one-half the thickness of the front and rear walls, so as to secure the front and rear walls to the side, end, or party walls; and all stone used for the facing of any building, except where built with alternate headers and stretchers, as hereinbefore set forth, shall be strongly anchored with iron anchors in each stone, and all such anchors shall be let into the stone at least one inch. The side, end, or party walls shall be anchored at each tier of beams, at intervals of not more than eight feet apart, with good, strong, wrought-iron anchors, one-half inch by one inch, well built into the side walls and well fastened to the side of the beams by two nails, made of wrought iron, at least one-fourth of an inch in diameter; and where the beams are supported by girders, the ends of the beams resting on the girder shall be butted together end to end, and strapped by wrought iron straps of the same size, and at the same distance apart, and in the same beam as the wall anchors, and shall be well fastened.

Walls to be coped when not corniced.

§ 13. All side or party and front or rear walls, not corniced, and where no gutter is required on any building, over fifteen feet high, shall be built up and extended at least twelve inches above the roof, and shall be coped with stone or iron, provided that, where partition walls are carried up, the partition walls may be carried up above the roofing and coped with some fire-proof material, or shall be carried up to the under-side of the roof-planking.........If a French or Mansard roof shall be placed upon any building, except a wooden building, over three stories in height, exclusive of the said roof, the same shall be constructed fire-proof; and if a French or Mansard roof shall be placed upon more than one side of any building, except a wooden building, located on the street corner, the same shall be constructed fire-proof throughout

Mansard roofs to be fire-proof.

§ 14. All iron beams or girders used to span openings Iron beams or girders, span of same and bearings required. over six feet in width, and not more than twelve feet in width, upon which a wall rests, shall have a bearing of at least twelve inches at each end by the thickness of the wall to be supported, and for every additional foot of span over and above the said twelve feet, if the supports are iron or solid cut stone, the bearing shall be increased half an inch at each end; but if supported on the ends by walls or piers built of brick or stone, if the opening is over twelve feet and not more than eighteen feet, the bearing shall be increased four inches at each end, by the thickness of the wall to be supported; and if the space is over eighteen feet and not more than twenty-five feet, then the bearing shall be at least twenty inches at each end by the thickness of the wall to be supported; and for every additional five feet or part thereof that the space shall be increased, the bearing shall be increased an additional four inches at each end by the thickness of the wall to be supported. And on the front of any building where the supports are of iron or Supports of stone or iron. solid cut stone, they shall be at least sixteen inches on the face and the width of the thickness of the wall to be supported, and shall, when supported at the ends by brick walls or piers, rest upon a cut granite base block, at least twelve inches thick, by the full size of the bearing; and in case the opening is less than twelve feet, the granite block may be six inches in thickness, by the whole size of the bearing; and all iron beams or Thickness of iron beams. girders used in any buildings shall be, throughout, of a thickness not less than the thickness of the wall to be supported. All iron beams or girders used to span Tie-rods. openings more than eight feet in width, and upon which a wall rests, shall have wrought-iron tie-rods of sufficient strength, well fastened at each end of the beam or girder,

and shall have cast-iron shoes on the upper side, to answer
for the skew-back of a brick or cut-stone arch, which
said arch shall always be turned over the same, and the
arch shall in no case be less than twelve inches in height,
by the width of the wall to be supported, and the shoes
shall be made strong enough to resist the pressure of
the arch in all cases. Cut-stone or hard-brick arches,
with two wrought iron tie-rods of sufficient strength,
may be turned over any opening less than thirty feet,
provided they have skew-backs of cut stone or cast or
wrought iron, with which the bars or tension-rods shall
be properly secured by heavy wrought iron washers,
necks, and heads of wrought iron, properly secured to
the skew-backs. The above clause is intended to meet
cases where the arch has not abutments of sufficient size
to resist its thrust. All lintels hereafter placed over
openings in the front, rear, or side of a building, or re-
turned over a corner opening, when supported by brick
piers or iron or stone columns, shall be of iron, and of
the full breadth of the wall to be supported, and shall
have a brick arch of sufficient thickness, with skew-backs
and tie-rods of sufficient strength to support the super-
incumbent lateral weight, independent of the cast-iron
lintel. In all buildings hereafter to be erected or altered,
where any iron column or columns are used to support
a wall or part thereof, whether the same be an exterior
or interior wall, except a wall fronting on a street, the
said column or columns shall be constructed as follows:
There shall be a double column, that is, an outer and
inner column, and the inner column shall be of sufficient
strength to sustain safely the weight to be imposed upon
both the outer and inner column; and the outer column
shall be made of sufficient size to allow a space of at
least one inch between it and the inner column, which

space shall be solidly filled with plaster of Paris, or some other non-conducting material; and all iron beams, girders, lintels, or columns, before the same are used in any building, shall have the maximum weight which they will safely sustain stamped, cast, or properly marked in a conspicuous place thereon by the founder or manufacturer of the same, and shall be made of the best materials and in the best manner.

Bearing weight of iron beams, columns, etc,. to be marked thereon.

§ 15. All openings for doors and windows in all buildings, except as otherwise provided, shall have a good and sufficient arch of stone or brick, well built and keyed, and with good and sufficient abutments, or a lintel of stone or iron, as follows: For an opening not more than four feet in width, the lintel shall not be less than eight inches in height; and for an opening not more than six feet in width, the lintel shall not be less than twelve inches in height; and for an opening exceeding six feet in width, and not more than eight feet in width, the lintel shall be of iron or stone, and of the full thickness of the wall to be supported; and every such opening six feet or less in width in all walls shall be at least one-third the thickness of the walls on which it rests, and shall have a bearing at each end not less than four inches on the walls; and on the inside of all openings in which the lintel shall be less than the thickness of the wall to be supported, there shall be a good timber lintel on the inside of the other lintels, which shall rest at each end not more than four inches on any wall, and shall be chamfered at each end, and shall have a double rolock arch turned over said timber lintel. . . . .

Openings for doors and windows.

Lintels and arches over same, and how set.

Fire-proof
doors and
shutters,
on what
class of
buildings
required.

§ 16. All stores or storehouses, or other buildings which are more than two stories or above twenty-five feet in height above the curb level, already erected, or that may hereafter be built in said city, except dwelling-houses, school-houses, or churches, shall have doors, blinds, or shutters made of fire-proof metal, on every window and entrance where the same do not open on a street. When in any such building the shutters, blinds, or doors cannot be put on the outside of such door or window they shall be put on the inside, and if placed on the inside shall be hung upon an iron frame independent of the wood-work of the window-frame or door;

When to be
closed.

and every such door, blind, or shutter shall be closed upon the completion of the business of each day by the occupant having the use or control of the same; and all fire-proof shutters or blinds that now are or may hereafter be put upon the front or sides of any building on the street fronts, must be so constructed that they can be closed and opened from the outside above the first story.

Wooden
furring.

§ 17. . . . . No wooden furring or lath shall be placed against any flue, metal pipe or pipes used to convey heated air or steam in any building; and when any wall shall hereafter be furred or lathed with wood, there shall be a strip of iron lath at least sixteen inches in width placed on the under side of the tier of beams running into the said wall, and extending at least one half inch into the horizontal joint of the brick wall, so as to prevent fire from extending from one floor

Hearths,
how
supported.

to another . . . : . . . . All hearths shall be supported by arches of stone or brick, and no chimney in buildings already erected or hereafter to be built shall be cut off below in whole or in part and supported by wood, but shall be wholly supported by stone, brick, or iron..........

§ 19. .... Every trimmer or header more than four feet long, used in any building except a dwelling, shall be hung in stirrup irons of suitable thickness for the size of the timbers. .... In every building already erected, or hereafter to be built, the floors shall be of sufficient strength to bear the weight to be imposed upon them exclusive of the weight of the materials used in their construction; and in all storehouses, the weight that each floor will safely sustain upon each superficial foot shall be estimated by the owner thereof, and posted in a conspicuous place on each floor thereof; and the weight that may be placed upon either of the floors of the said building or buildings shall be safely distributed thereon............

Trimmers or headers, length of same, and how hung.

Strength of floors.

§ 20. In all buildings, every floor shall be of sufficient strength in all its parts to bear safely upon every superficial foot of its surface seventy-five pounds; and if used as a place of public assembly, one hundred and twenty pounds; and if used as a store, factory, warehouse, or for any other manufacturing or commercial purposes, from one hundred and fifty to five hundred pounds and upwards; and every floor shall be of sufficient strength to bear safely the weights aforesaid, in addition to the weight of the materials of which the floor is composed; and every column, post, or other vertical support shall be of sufficient strength to bear safely the weight of the portion of each and every floor depending upon it for support, in addition to the weight required as above to be supported safely upon said portions of said floors. In all calculations for the strength

Bearing weight of floors.

of materials to be used in any building, the propor-

Calculations for strength of materials. tion between the safe weight and the breaking weight shall be as one to three for all beams, girders, and other pieces subjected to a cross-strain, and shall be as one to six for all posts, columns, and other vertical supports, and for all tie-rods, tie-beams, and other pieces subjected to a tensile strain. And the requisite dimensions of each piece of material is to be ascertained by computation by the rules given by Tredgold, Hodgkinson, Barlow, or the treatises of other authors now or hereafter used at the United States Military Academy at West Point on the strength of materials, using for constants in the rules only such numbers as have been deduced from experiments on materials of like kind with that proposed to be used. Before any iron column,

Iron beams, girders, etc., to be tested before being used. beam, lintel, or girder, intended to span an opening over eight feet in length, and intended to support a wall built of stone or brick, or any floor or part thereof, in any building hereafter erected or altered, in the City of New York, shall be used for that purpose, the manufacturer or founder thereof shall have the same tested by actual weight or pressure thereon, under the direction and supervision of an inspector of the department of buildings, authorized by the superintendent thereof (who shall be previously notified of the time when and place where the said test will be made by the person or

Bearing weight to be marked thereon. persons having the said columns, beams, lintels, or girders so tested), whose duty it shall be to have the weight which each of the said columns, beams, lintels, or girders will safely sustain properly stamped or marked in a conspicuous place thereon by the said manufacturer or founder thereof, and no greater weight shall be put or placed upon any said column, beam, lintel, or girder than the same shall be so-marked as being capable of sustaining; and in case any said column, beam, girder, or lintel shall be rejected by said inspector as unfit to be used, the same shall not be used in, upon, or about any building or part thereof. All iron-work used in any building or part thereof hereafter to be erected or altered shall be of the best material and made in the best manner. *Iron to be of the best quality.*

§ 21. In all fire-proof buildings hereafter to be con-*Fire-proof buildings.* structed, where brick walls, with wrought-iron beams or cast or wrought-iron columns with wrought-iron beams, are used in the interior, the following rules must be observed:

1. All metal columns shall be planed true and *Metal columns.* smooth at both ends, and shall rest on cast-iron bedplates, and have cast-iron caps, which shall also be planed true. If brick arches are used between the beams the arches shall have a rise of at least an inch and a quarter to each foot of space between the beams.

2. Under the ends of all the iron beams, where they *Stone template.* rest on the walls, a stone template must be built into the walls; said templates to be eight inches wide in twelve-inch walls, and in all walls of greater thickness to be in width not less than four inches less than the width of said walls, and not to be, in any case, less than four inches in thickness and eighteen inches long.

3. All arches shall be at least four inches thick. *Arches.* Arches over four feet span shall be increased in thickness toward the haunches by additions of four inches in thickness of brick; the first additional thickness shall commence at two and a half feet from the centre of the span, the second addition at six and a half feet from the centre of the span, and the thickness shall be

increased thence four inches for every additional four feet of span towards the haunches.

4. The said brick arches shall be laid to a line on the centres, with a close joint, and the bricks shall be well wet, and the joints well filled with cement mortar, in proportions of not more than two of sand to one of ce- *Arches to be grouted.* ment, by measure. The arches shall be well grouted and pinned or chinked with slate and keyed.

Cornices and gutters to be fire-proof, and how set. § 22. All exterior cornices and gutters of all buildings hereafter to be erected or built, shall be of some fire-proof material, and in every case the greatest weight of stone, iron, or other materials of which the cornice shall be constructed, shall be on the inside of the outer line of the wall on which the cornice shall rest, in the proportion of three of wall to two of cornice in weight, allowance being made for the excess of leverage produced by the projection of the cornice beyond the face of the wall; and all fire-proof cornices shall be well secured to the walls with iron anchors, independent of any wood-work; and in all cases the walls shall be carried up to the planking of the roof, and where the cornice projects above the roof, the wall shall be carried up to the top of the cornices, and the party wall shall, in all cases, extend up above the planking of the cornice, and be coped with some fire- *Metallic leaders, and how set.* proof material......... All buildings shall be kept provided with proper metallic leaders for conducting the water from the roof to the ground, sewer or street gutter, in such manner as shall protect the walls and foundations from damage; and in no case shall the water from the said leaders be allowed to flow upon the side-walk, but shall be conducted by drain pipe or pipes to the street gutter or sewer.

§ 23. The planking and sheathing of the roof of every building erected or built as aforesaid, shall in no case be extended across the front, rear, side, end, or party wall thereof, and every such building, and the tops and sides of every dormer-window thereon, shall be covered and roofed with slate, tin, zinc, copper, or iron, or such other equally fire-proof roofing as the superintendent of buildings, under his certificate, may authorize, and the outside of the frame of every dormer-window hereafter placed upon any building as aforesaid, shall be made of some fire-proof material.......... All buildings shall have scuttle-frames and covers, or bulkheads and doors, made of or covered with some fire-proof material, and all scuttles shall have stationary iron ladders leading to the same, and all such scuttles or ladders shall be kept so as to be ready for use at all times, and all scuttles shall not be less in size than two by three feet. .... All skylights more than three square feet, placed in any building, the sash and frames thereof shall be constructed of fire-proof materials.

§ 28. In any building hereafter erected more than three stories in height occupied by or built to be occupied by three and not more than six families above the first story, in which the cellar is to be used for the purpose of storing coal, wood, or other articles, the floor above the cellar with the stairs leading thereto, if the stairs lead from the inside of the building, shall be constructed fire-proof; and where the lower part thereof is to be used for business purposes of any kind, the first floor, if there is a cellar below, and the ceiling above the store floor shall be constructed fire-proof, and the hall partition and partitions from front to rear, from

[margin notes:] Fire-proof roofing. Scuttles and ladders. Skylights and frames. Fire-proof floor and stairs in tenement houses.

the cellar to the top of the second-story beams, shall be built of brick; and in no case shall a front and rear tenement house be erected on the same lot unless there shall be an open space of at least twenty-eight feet the

whole width of the lot between the same. All the window openings of all rear buildings and all the rear window openings of all buildings mentioned in this section shall be provided with fire-proof blinds...... ..

Any dwelling-house now erected or that may hereafter be erected in said city more than two stories in height, occupied by or built to be occupied by two or more families on any one of the floors above the first story, and all dwelling-houses now erected or that may hereafter be erected more than three stories in height, occupied by or built to be occupied by three or more families above the first story, and any building already erected or that may hereafter be erected more than two stories in height, occupied as or built to be occupied as a hotel, boarding or lodging house, factory, mill, offices, manufactory, or workshops, in which operatives are employed in any of the stories above the first story, shall be provided with such fire-escapes, alarms, doors, and ventilators as shall be directed and approved by the said su-

perintendent of buildings. And the owner or owners of any building upon which any fire-escapes may now be or may hereafter be erected, shall keep the same in good repair and well painted, and no person shall at any time place any encumbrance of any kind whatsoever upon any said fire-escapes now erected or that may hereafter be erected in said city.

§ 30. Before the erection, construction, alteration, or repair of any building or part of a building is commenced, the owner, architect, or builder shall notify the superintendent of the department of buildings, and shall submit to said superintendent a detailed statement in writing of the specification and also a copy of the plans for such erection, construction, alteration, or repair, and a record of said statement and copy of the plans shall be kept in the office of the said department of buildings; and the erection, construction, alteration, or repair of the said building, or any part thereof, shall not be commenced or proceeded with until the said specification and plans shall have been approved by the said superintendent of buildings.

§ 32. .... Any and all persons who shall violate *Penalties.* any of the provisions of this act, or fail to comply therewith, or any requirement thereof, or shall in any manner be liable therefor, shall severally, for each and every such violation and non-compliance, respectively forfeit and pay a penalty in the sum of fifty dollars; and any and all persons who shall violate any of the provisions of this act, or who may be employed or assist therein, or who shall be liable therefor, shall severally, for every such violation not removed or requirement not complied with, within ten days after notice thereof shall be given to him or them respectively, forfeit and pay an additional penalty in the sum of fifty dollars, for the recovery of which said penalties, or either of them, an action may be brought in any court of competent jurisdiction; .... and the *Penalties may be remitted.* superintendent of buildings is hereby authorized, in his discretion, good and sufficient cause being shown therefor, to remit any fine or fines, penalty or penalties, which any person or persons may have incurred, or may hereafter incur, under any of the provisions of this act.

§ 35. Any and all persons who, after having been
personally served with the notice of violation as hereinbefore prescribed, shall fail to comply therewith,
and shall continue to violate any of the several provisions of this act, or who shall be accessory thereto,
shall, in addition to the other penalty or penalties in
this act provided, be deemed guilty of a misdemeanor,
and, upon a complaint being made by the superintendent of buildings, before any police justice or any
court of criminal jurisdiction within the City of New
York, shall be arrested and held to bail by said justice or said court, and, upon conviction of such of-
fence, shall be fined in a sum not exceeding two hundred and fifty dollars, or may be imprisoned for a term
not to exceed six months ; said fine or imprisonment to
be imposed in the discretion of the judge, justice, or
court by whom said person so arrested and held to bail
shall be tried.....................................

[The foregoing extracts from the Building Law covers
all that relates to iron work. The law itself is, of
course, a local one of the City of New York ; but its
provisions, so far as iron work is concerned, are sound,
common sense requirements, which would be well to
carry out without regard to locality.]

DEPARTMENT OF BUILDINGS,

IN THE CITY OF NEW YORK.

OFFICE OF SUPERINTENDENT,

New York,................ .....,187

Sir :

When you desire to have either Iron Beams, Lintels, or Girders tested, agreeably to the requirements of Section 20, Chapter 625, Laws of 1871, you will please fill out this blank, in order that the proper pressure requisite for the test may be readily ascertained. This blank must in all cases be filled out before the test is made. Respectfully yours,

....

Superintendent of Buildings.

Please test for (name)..................(business)..........
.......................(place of business).............(description of articles to be tested)........................... ....
....
to be used in building No................. ....to be tested to sustain
... .tons of 2,000 pounds.   (Owner).....................
                                          (Signature) ....
                                          (Business)...................

FOR A BEAM, LINTEL, OR GIRDER.

1st. Is the weight to be sustained at rest, or subject to vibration ? Ans.....
2d. What is the distance in the clear between supports?....feet....inches.
3d. What is the beam, lintel, or girder supported by ? Iron or stone columns, brick wall or piers?................................ ....
....
4th. What are the bearings on the wall, at each end ?.........inches.
5th. What is the full length of the beam, lintel, or girder?....feet....inches long.

IF TO SUSTAIN A WALL, BRICK ARCHES, OR ANY OTHER BRICK OR STONE WORK,

Give the thickness of the walls of each story and height of each story, also the full height of the wall from the lintel................................

	Inches thick.	Height of each story.	Weight per foot.	Total pounds per story.	Tons weight.
1st story wall	...... in.	...... feet	......lbs.	.... lbs.	....
2d " "	.... "	.... "	.... "	.... "	....
3d " "	.... "	.... "	.... "	.... "	....
4th " "	. .... "	.... "	.... "	.... "	....
5th " "	.... "	.... "	.... "	.... "	....
6th " "	.... "	.... "	.... "	.... "	....
7th " "	.... "	.... "	.... "	.... "	....
8th " "	.... "	.... "	.... "	.... "	....

Total Tons Weight.

Estimate the weight of wall per foot in height of wall, as follows :

8 in. brick wall, weight per foot, 77 lbs.		Brown Stone, 4 inches.......... 57 lbs.
12 " " " 115 "		" " 8 " ....114 "
16 " " " 153 "		" " 12 " ....170 "
20 " " " 192 "		Granite " per foot.... 166 "
24 " " " 230 "		White Marble " ....168 "

What is the full height of the wall from the bearing of the beam, lintel, or girder ?.............feet............. ...inches.

If this weight is not equally distributed, double it.

Should it sustain a chimney, or any other weight, add the additional weight in all cases.

Deduct for windows only half weight; that is, take out of the weight imposed on beam, lintel, or girder, but half the actual space which the windows will occupy. *Deduction*.............

....

....

Total Tons imposed....

NOTE.—Should a pier rest on or about the middle of beam, lintel, or girder, the weight must not be considered to be equally distributed. In computing the weight of a brick arch, estimate a four-inch arch as equal in weight to an eight-inch thick wall, and an eight-inch arch as equal in weight to a twelve-inch thick wall, on a straight line. This additional weight is to make allowance for the weight of material required to fill up on a level with the crown of the arch. Make additional allowance for any material placed above the crown of the arch.

IF TO SUSTAIN FLOORS, GIVE | Pounds.

Size of floors .....feet......inches wide × ....feet .....inches long =..

Number of floors ?...

What is the area of floor surface ?........feet... ....inches.

What is the weight of floor, per superficial foot ?.....pounds per foot =.............lbs................................... ....

(See *)

Should this weight of floors rest on a girder or girders—which rest either directly on the iron beam, lintel, or girder, or on the wall above, which it sustains—the weight must in every case be doubled, as the weight is considered central and other than equally distributed.

............pounds from above, doubled, =.................... ....

....

Should the weight of floors be sustained on beams resting on the iron beam, lintel, or girder, or on the wall above, which it sustains, equally distributed over its length, it does not require to be doubled.

....

....

IF FLAT ROOF SURFACE.

......feet.....inches long × ... .feet.......inches wide,

=.......feet.....inches, at 90 pounds per foot =..... ....

* For Mansard Roof, additional calculation will have to be made for the weight imposed.
* Should the Iron Beam, Lintel, or Girder, sustain Tanks or any other weight, the calculation to be made on this sheet.

....

....

....

Total pounds............................ ....

For Tenement Houses, compute the weight per foot floor surface... 100 lbs.
Dry Goods House.. 310 "
Flour Store.. 350 "
Public Assemblies... 180 "
Roof, including snow.. 90 "
Hardware Store.................................... from 350 to 600 "

* For cast-iron arch beams or girders with wrought-iron tension rods, calculate the maximum strain when the pressure or weight of test is applied on the middle of beam or girder, not to exceed five tons per square inch of tension rod, or equal to ten tons distributed.

REPORT OF INSPECTOR.

NEW YORK, ..., 187

To the Superintendent of Buildings;

I respectfully report that the iron girders, beams, and lintels, described in the foregoing application, were practically tested by me with the following result:

Tested to..........Tons......., Deflected............inches.
"   .... " ...., " .... "
"   .... " ...., " .... "

Permanent Set.

I hereby certify, by the foregoing test, that the above .... is .... sufficient to bear the weight to be imposed thereon, agreeably to the requirements of the annexed application, and having .... approved of the same, I have .... caused the mark of the Department to be placed thereon.

....
Inspector of Iron Work.

BLANK FORM OF REQUEST FOR ESTIMATE.

OFFICE OF
.................., ARCHITECT,
No...., ... STREET.
............................., 187

To.........................

DEAR SIR: You are hereby invited to submit proposals for the Iron Work required for building ....

Plans and Specifications are now ready at this office. Bids will be received until the ...... day of ......, at 12 m.

Yours truly,

....

BLANK FORM OF PROPOSAL.

OFFICE OF
..........................,
IRON CONTRACTOR AND MANUFACTURER,
No. ...., ... STREET.
............................., 187

To.........................

DEAR SIR: I (or we) hereby propose to do all the Iron Work required by the Specification of Iron Work and Plans, for the building to be erected No......, ... Street, for the sum of .... dollars, $..........

Respectfully,

BLANK FORM OF CONTRACT.

ARTICLES OF AGREEMENT made this .... day of ...., in the year one thousand eight hundred and..............
BETWEEN ....
of the first part, and,...
of the second part.

First. The said *part* of the second part *do* hereby for ...... heirs, executors and administrators, covenant, promise and agree to and with the said *part* of the first part ...... executors, administrators or assigns, that ...... the said *part* of the second part, ... executors or administrators, shall and will, for the consideration hereinafter mentioned, on or before the .... day of .... next, well and sufficiently erect and finish the IRON WORK of the building to be erected, built and completed on the land of the *part* of the first part, known as lot No. ...., ... Street, in the city of ...., agreeably to the Drawings and Specification made by ...., Architect, and signed by the said parties and hereunto annexed, within the time aforesaid, in a good, workmanlike and substantial manner, to the satisfaction, and under the direction of the said Architect, to be testified by a writing or certificate under the hand of the said Architect, and also shall and will find and provide such good, proper and sufficient materials, of all kinds whatsoever, as shall be proper and sufficient for the completing and finishing all the IRON WORK and other works of the said building mentioned in the Iron Specification for the sum of .... dollars. And the said *part* of the first part, *do* hereby for ...... heirs, executors and administrators, covenant, promise and agree, to and with the said *part* of the second part, ... executors and administrators, that ...... the said *part* of the first part, ... executors or administrators, shall and will, in consideration of the covenants and agreements being strictly performed and kept by the said *part* of the second part, as specified, well and truly pay, or cause to be paid unto the said *part* of the second part, ... executors, administrators or assigns, the sum of .... Dollars, lawful money of the United States of America, in manner following :

When				
"	the sum of	Dollars.		
"	the sum of	Dollars.		
"	the sum of	Dollars.		
	Total,........$			

PROVIDED, that in each of the said cases, a certificate shall be obtained and signed by the said Architect.

AND IT IS HEREBY FURTHER AGREED BY AND BETWEEN THE SAID
PARTIES:

First. The Specification and the Drawings are intended to co-operate, so
that all work mentioned in the Specification is to be executed the same as set
forth in the Drawings, to the true meaning and intention of the said Drawings and Specification, without any extra charge whatsoever.

Second. The Contractor, at his own proper cost and charges, is to provide
all manner of materials and labor, scaffolding, implements, moulds, models
and cartage of every description, for the due performance of the several
erections.

Third. Should the Owner, at any time during the progress of the said
Iron Work, request any alteration, deviation, additions or omissions, from the
said contract, he shall be at liberty to do so, and the same shall in no way
affect or make void the contract, but will be added to, or deducted from, the
amount of the contract, as the case may be, by a fair and reasonable valua-
tion.

Fourth. Should the Contractor, at any time during the progress of the
said works, refuse or neglect to supply a sufficiency of materials or workmen,
the Owner shall have the power to provide materials and workmen, after
three days' notice in writing being given, to finish the said works, and the
expense shall be deducted from the amount of the contract.

Fifth. Should any dispute arise respecting the true construction or meaning of the Drawings or Specification, the same shall be decided by the said
Architect, and his decision shall be final and conclusive; but should any
dispute arise respecting the true value of the extra work, or of the works
omitted, the same shall be valued by two competent persons—one employed
by the Owner, and the other by the Contractor—and those two shall have
power to name an umpire, whose decision shall be binding on all parties.

Sixth. The Owner shall not, in any manner, be answerable or accountable
for any loss or damage that shall or may happen to the said works, or any
part or parts thereof respectively, or for any of the materials or other things
used and employed in finishing and completing the same (loss or damage by
fire excepted). The Owner shall keep the said building insured, and be
responsible for all loss or damage by fire.

IN WITNESS WHEREOF, the said parties to these presents have hereunto
set their hands and seals, the day and year above written.

Witnesses: [L. S.]

                                                          [L. S.]

FOUNDING.

The following general remarks on casting, moulding, etc., is
intended more particularly for the benefit of those who may
have invested money in a foundry, and yet who know little or
nothing of the practical workings of such a place.

Iron is generally melted in a Cupola-furnace. The cupola,

as ordinarily constructed, consists of four legs and a cast-iron circular or elliptical flanged plate, which is laid on the legs, and has an opening in the centre, to which swinging doors are fitted. A wrought-iron exterior shell rises from the plate, to a height of from 10 to 18 feet, and is lined with fire brick. The inside diameter is from 3 to 8 feet. The chimney or flue is also of boiler iron, lined with fire brick. The cupola is capable of melting from 3 to 20 tons of pig iron per hour. Near the bottom is an opening in the brick lining, through which the melted iron runs when the furnace is tapped. A little higher up two isinglass peep-holes are provided for showing the state of combustion and position of the coal, etc., inside. To create a draught, a current of air is forced in at the sides, through tubes called tuyeres, by a blowing machine.

a a, Tuyeres. b b, Small isinglass windows. c, Ladle for receiving the melted metal. d d d, Columns of support. The small upright rods support the hinged floor, and stand in the pit below the cupola.

EXTERIOR OF THE LOWER PART OF A CUPOLA.

In the upper part of the back of the cupola is the opening for receiving the charges. When the cupola is to be used, the swing doors are raised and held up by upright bars. The doors are then covered on the inside with sand, to the depth of about six

inches. The charging is done by placing a sufficient quantity
of kindling wood upon the floor, and above this a layer of the
best anthracite coal in large lumps, and in sufficient quantity to
fill up to the height of several inches above the line of tuyeres,
after it has well settled and the wood has burned away. This
precaution must be carefully observed, because if the charge of
iron above the coal should come down to a level with the entrance of the blast, combustion would be checked, the metal
become chilled, the process stopped, and the dumping of the
charge necessitated. Upon the layer of coal thus carefully deposited, one of pig iron is placed, varying in quantity from
1,000 to 5,000 lbs., according to the size of the cupola and
to the rapidity with which it is proposed to effect the melting ;
and upon this another layer of coal is deposited, and afterwards
succeeding layers of iron and coal. The pig is broken into pieces
from ten to fifteen inches in length before it is charged.
Fluxes are added where occasion requires, according to the
judgment of the melter, pounded marble or limestone being
most frequently employed. The wood is usually ignited when
the first layer of coal is deposited, and in an hour to an hour
and a half the furnace may be tapped. When the charging is
being done, behind the opening through which the molten iron
is to be let or tapped a lump of coal is so placed that the opening may be rammed full of refractory material, preferably
moulding sand. This tap-hole is 1½ or 2 inches wide, and is
formed by placing a tapered round bar in the place where the
hole is to be, ramming the sand tightly around it and removing
it as soon as the hole is filled up. The sand filling is usually as
thick as the cupola lining—say six to nine inches—and this part
is called the breast of the cupola. The tap-hole is closed by a
stopper, made of loam, which is worked in the hand until it
assumes a certain degree of tenacity ; a round ball of it is then
fastened on the end of a stick of wood, provided with a disc
of iron, which, being previously wet, is then pressed into the

tap-hole. This stopper is removed and replaced as often as required during the operation of tapping.

On the inside of the cupola the space just above the tuyeres has the shape of a cone, which has the effect to hold the contents in such a relation to the blast as is best calculated to make it the most effectual. Being larger toward the bottom than at the top, it works hotter than if made with parallel sides, and also has the advantage of lasting longer, as the melted iron which is apt to cut the fire-brick, does not run down along the brick. Slag, or scoria, more or less sticks to the lining and gradually fills up the cupola, and finally compelling the brick to be torn out and new put in. The melting and pouring is done usually each afternoon. The molten iron after being let out through the runner is caught into pots and ladles of various sizes and capacities, to be carried by hand or lifted by cranes from the furnace to the moulds. The ladles are covered by a coating of clay, put on every cast anew. The large pots are always made of wrought iron, the smaller ones may be either cast or wrought iron. The sand bottom of the furnace is made sloping, so as to admit of discharging the last portions of the iron. It will be understood that in melting iron, wood, and coal all together, the iron being the heaviest works through and falls to the bottom, the purest iron being at the lowest point, and the dross, impurities, etc., being on top. The dross from the iron after being received in the ladles, rises to the top and is skimmed off, or held back when the moulder is pouring into the moulds. The melting point of cast iron varies. Scotch pig melts at a somewhat lower temperature than American pig, because of its larger quantity of carbon. It is a common practice among founders to melt different brands of iron together, to give the mixture desired characteristics which they do not possess separately. The cupola has the advantage of melting iron cheaper than any other furnace, and enables a large or small quantity to be melted. After the iron has all been tapped out,

12

the bars under the swing doors are knocked away and the remaining contents of the furnace dumped, the debris falling into the brick pit, a stream of water through a hose played upon it for a while, and the following day shovelled out and assorted.

MOULDING.

The floor of the foundry, for a depth varying from 3 to 8 feet, is made up of moulding sand. The first matter to receive attention in moulding is the selection and proper treatment of the sand; for it is only by the use of sand possessing certain properties, that the formation and retention of a smooth and well defined cavity can be produced, having at the same time sufficient porosity to allow of the escape of air and gases which are generated during the pouring of the metal. It must possess, in a certain degree, the nature of a plastic or adhesive substance. The various kinds of good moulding sand have been found to be of an almost uniform chemical composition, varying in grain or the aggregate form only. It contains between 93 and 96 parts of silex or grains of sand, and from 3 to 6 parts of clay, and a little oxide of iron, in each 100 parts. It has in its green state a yellowish earthy color, balls easily on being squeezed in the hand, and, if sufficiently fine, assumes the finest impressions of the skin without adhering to it. Sand is more or less porous, and very refractory, so that the hot metals do not melt or bake it—two qualities of great importance in the successful operations of the business. In practice, the different classes of castings require different kinds of sand; for one kind the sand is to be porous, open, and is still to be adhesive; for another class it is to be very adhesive and fine, almost free of grit, to make itself conform to the minutest parts of the pattern embedded in it. Enough moisture must be present in the sand to produce a proper degree of adhesion, but the quantity must be as small as possible, for too much would produce an amount of vapor when the molten metal is poured that would injure or destroy the mould. The cost of sand is not an item of much consequence after the first supply is obtained, as it is used over and over many times. It has often to be transported considerable distances, as it is not found in every locality where common sand exists. The moulding sand which is used in New York City is principally obtained in New Jersey and in the vicinity of Albany, N. Y. To work successfully in green sand (as it is called), it is almost absolutely necessary to divide the articles of manufacture. The moulder who has been trained to small articles is hardly able to do heavy work; and those moulders who have been used to heavy articles cannot compete with moulders of light castings. The sand suitable for columns, beams, etc., is not fit for leaves of capitals, cornices, etc. There needs to be a separate shop, and separate hands, particular sand for light and heavy work respectively.

The tools used by moulders are various, consisting of trowels from the size of a small mason's trowel, down to a very small one; and tools for polishing and cleaning surfaces, together with rammers, pointed and round. Besides the tools here enumerated, the moulder has short-handled light shovels, for filling boxes and for working the sand; sieves of various sizes or meshes, and a riddle for filling the flask; small bellows for blowing dry, loose sand from the moulds, and parting sand from the patterns, etc. The moulder needs an iron water pot; two or more linen bags for coal dust, black lead, etc., a piece of rope for tufts, and iron or brass piercers or prickers.

Architectural casting is mainly done in green sand. Many articles require a combination of dry sand and green—dry sand for cores. Cores are especially used for forming vacancies in castings, which cannot be successfully formed by the pattern. Core sand should be coarse and very porous, such as white sand from the sea-shore. It is mixed with flour, sour beer, etc., forming a paste, and baked hard in the core oven. Fresh sand must

be used in each cast; old sand, burned sand, or sand mixed
with coal, cannot be employed for this purpose. The casting
of a hollow column is an example of mixed moulding in green
and dried sand. The outer part of the mould is made in a flask
of two parts with green sand, from a solid pattern of the column.
A dry sand core somewhat longer than the mould is then placed
in the axis of the hollow mould, its extremities resting upon the
sand beyond. The thickness of the walls of the column will of
course be in inverse proportion to the size of the core. The
management of cores is a matter which requires some ingenuity.
A caution not to be neglected is that cores are never to be put
into a green sand mould until the very latest moment before
casting.

Long or thin cores, whether in green sand or dry, are stiffened
by arbors, or small rods of iron, which are moistened with clay
water. Such wires or rods are buried in the core and recovered
when the casting is cleansed from its adhering sand. If cores
are too long to bear their own weight and the pressure of the
metal, they are to be supported by wires or chaplets, so that the
cores will be kept at the right distance from the mould.

Coal-dust, black-lead, and anthracite dust, are simple means
of blackening the mould by mixing it with sand. If hot metal
is allowed to be in immediate contact with some kinds
of fresh sand, the sand will partly melt; or if the sand is
coarse, the hot metal will penetrate into the spaces between the
grains, and the casting in consequence will be rough. Blackening, or a coating of carbon, will prevent in a great measure the
burning of the sand, and consequent roughness of the casting;
but if used in too large quantities it is apt to fill the necessary
pores of the sand, and, as it is almost incombustible, will prevent the escape of gases from the hot metal, and consequently
cause unsound castings. Sharp outlines can never be expected
if too much coal is used, either mixed with the sand or dusted
on. In ornamental moulding it is not generally the strength of

the metal which is the most valuable, but it is the perfect representation of the pattern which is desirable. Sharp outlines and smooth castings are the main object.

The art of moulding would in itself fill a volume. It is a trade, and skilful workmen are plentiful. After all, the general appearance of finished architectural iron work does not depend so much on the surfaces of the castings, as it does on the patterns from which the castings are made—the boldness of outlines, the artistictness of the carvings—things beyond the control of the moulder. The moulding of these patterns is generally simple, there are but few complicated forms, and therefore there is no reason why well-finished castings of uniform thickness should not be turned out. Indeed the most elegant castings, surpassed nowhere in the world, are now made for architectural purposes. There is a great advantage in well-finished patterns. If the patterns are perfect the castings will be good.

The moulds are generally formed in a frame similar to a box, without top or bottom, and having traverses or bars running across on the inside. These boxes are technically called "Flasks," and they enclose the sand which is filled around the pattern. A flask is made in two or more parts, the top portion being called the cope, and the bottom the drag. On each side of the flask are two or more hooks fitting to eyes which serve to connect the two parts of the flask as firmly as possible, to prevent a separation or the lifting of the upper box. Pins are also arranged in the sides, so that the boxes can be lifted apart and brought back again to exactly the same position. On each box are four snugs or handles for lifting and carrying. Flasks are made as rough as possible inside, for it is by adhesion chiefly that the sand remains in the box. The adhesion of the sand is increased by driving nails into the traverses and sides of the box, of such length that the points project on the inside.

Flasks may be made of wood or of cast iron. Iron Flasks

are in the course of time the cheapest, as they are the strongest and most durable. Wooden flasks burn and leak, and never make correct castings; their pins never fit well and the wood is apt to warp. Iron flasks have the same general construction as wooden ones, with the addition of strong ears, by which they may be lifted with a crane. The adhesion of the sand is secured by nails being cast in the box, or its inner surface is covered with projections made by driving the piercer an inch or so into the sand before casting the box. The form of box is generally suited to the pattern, and must always be strong enough to resist the influence of the heavy weight of sand and iron. If the box gives way the sand will crack and drop out, spoiling the mould.

Large flasks are held together by clamps or dogs, and heavy weights to hold the cope down, many pieces, such as box columns, etc., being moulded in the floor. When the frame of the box is made of iron the traverses are often of wood. The interior of a flask is made wet, traverses and all, with a solution of strong loam or clay, put on by means of a brush.

The cost of flasks is a serious item. Expensive as they are to make, they are only worth the price of fire-wood and old iron under the auctioneer's hammer. Unless carefully guarded against, too many flasks will be made and too much iron tied up in weights, etc.

The hot metal is poured into the mould through a git, or gate, which is simply a tapering hole through the upper box. The hole is formed by setting in one or more wooden pins in the sand. The setting of these for gits is a nice point and requires considerable discrimination on the part of the moulder. The gits are to be very tapering and smooth, to allow an easy passage for the hot metal and prevent the washing down of loose sand. Holes and vents must be provided for the escape of air and gases. A powerful expansive force is applied to the interior of the mould when the hot metal is poured in, and the greatest precautions must be taken to have all the iron fastenings, as well as the sand tampings, strong enough to withstand the pressure.

The work of casting is the last business of the day. After casting, the small articles are removed the same afternoon, and the heavy pieces during the night time. After the sand is rubbed from them they are carried to an adjoining apartment to be chipped and otherwise finished. The flasks are piled up so as to be handy for the next day's work; and the sand, after receiving some water, is shovelled over, mixed, and thrown in heaps, where it remains during the night. If properly performed, the sand will be of a proper and uniform dampness the next morning. The expense of moulding is very variable. It is done by days' work and by piece work.

Castings are generally made of greater thickness in practice than the requirements of theory show as necessary. Much of the strength of a casting depends on the design. There should be as few abrupt bends, sharp angles, and sudden variations of thickness as possible, in order to obtain equal and uniform cooling and accord in the order and direction of crystallization, as it has been found from experience that wherever the order of crystallization is disturbed there will be found weakness. Increased thickness should not be considered as an equivalent to inferior iron, for in no material can it be said with greater truth that it is absolutely necessary to have quality as well as quantity.

WAR PRICES.

The following List-Prices of the Iron Founders' and the Engineers' Association of New York, in 1864–65, will be found valuable as a reference to the ruling high charges for labor and materials during war times, and for comparison between then and now:

PRICES

ADOPTED BY THE

IRON FOUNDERS OF NEW YORK AND NEIGH-BORING CITIES,

OCTOBER 1, 1864.

MACHINERY CASTINGS.

Ordinary Green Sand Castings.................. 7 cts. per lb. and upwards.
Dry Sand Castings............................. 8½ "   "   "   "
Loam Castings..... .... 9 "   "   "   "
Heavy Grate Bars.............................. 6 "   "   "   "
Light   "   "   .... 6½ "   "   "   "

Pattern making............................. $4 per day.

SHIP CASTINGS.

Ordinary Green Sand Castings.................. 7 cts. per lb. and upwards.

HOUSE WORK.

ROUND COLUMNS.

Not exceeding half-inch thick, with ordinary Cap and Base Plates.

3½ inches diameter ....$1.40 per foot.
4   "   "   .... 1.63   "
4½   "   "   .... 1.96   "
5   "   "   .... 2.33   "
6   "   "   .... 2.80   "
7   "   "   .... 3.50   "

COLUMNS.

Heavy Round Columns.......................... 7 cts. per lb. and upwards.
Corinthian Columns........................... 7 "   "   "
Box Columns .... 7 "   "   "
I Columns.................................... 7 "   "   "
⊥ Columns .... 7 "   "   "
Corinthian Capitals.......................... extra price.

LINTELS AND SILLS.

Box Lintels.................................. 7 cts. per lb. and upwards.
I   "   .... 7 "   "   "
⊥   "   .... 7 "   "   "
Door Sills................................... 7 "   "   "

GIRDERS AND BEAMS.

Vault Girders and Beams....................... 7 cts. per lb. and upwards.
工   "   .... 7 "   "   "
⊥   "   .... 7 "   "   "
Arch   "   .... 7 "   "   "
Wrought Iron Rods......................... 15 cts. and upwards.
Castings for Buildings not included in above list, 7 cts. per lb. and upwards.

RAILING CASTINGS.

Heavy Railing Castings......................... 7 cts. per lb. and upwards.
Light   "   "   .... 7¼ "   "   "
Cored Balusters......................... .... extra price.

RANGE AND FURNACE CASTINGS.

Range Castings.. 7¼ cts. per lb.
Furnace, Ship, Stove, and Hotel Range Castings............ 7 "   "

SEWING MACHINE CASTINGS...................7½ cts. per lb. and upwards.

HEAVY ANVIL BLOCKS, Buoy Weights, and Ballast may be made by special agreement at ten per cent. above cost.

N.B.—All the above prices are net cash.

WAR PRICES.

TARIFF OF THE ENGINEERS' ASSOCIATION.

New York, April 5, 1865.

Machinists, in Shop, or out, and on all Jobbing Work.... each per day, $4.25
Pattern Makers,....................................... "   "   4.50
Millwrights,.. "   "   4.25
Boiler Makers,.. "   "   4.25
Blacksmiths, man and helper at small fire, ... "   12.00
  do.   do.,   at large fire, ... from $15 to 20.00
Extra Helpers, ... each per day, 3.00
Laborers, ... "   "   2.50
Large Slide and Facing Lathes and Planer,............. "   "   15.00
  do.   Second Class, ... "   "   12.00
  do.   Third   do. .... "   "   10.00
  do.   Fourth do. .... "   "   8.00
Slotting Machines, First Class......................... "   "   15.00
  do.   do.   Second do .... "   "   12.00
  do.   do.   Third do .... "   "   10.00
Shaping Machines, First Class,....................... "   "   10.00
  do.   do.   Second do. .... "   "   8.00
Drilling Machines, Bolt Cutters, and other Similar Tools "   "   8.00
Boiler Iron, for Repairs, ...per lb.   .12
Rivets, ... "   .13
Bar Iron,... "   .10
Loam and Dry Sand Castings,......................... from 8 c. " to .10
Machinery Green Sand Castings, ... from 6 c. " to .08
Grate Bars,.. "   .06
Brass Castings..................................... from 60 c. " to .85
Copper Pipe, ... "   .85

EXHAUST TRAPS FOR STEAM PIPES.

A simple and effective apparatus placed at the top of the exhaust pipe, to catch and carry off the water which otherwise would fall on the roof or in the street below, is the Trap known as CONROW's PATENT, and manufactured and for sale by most of the principal Steam Fitters. Its use is very important on buildings having Iron Fronts; indeed, it is required on the top of every building in which steam is used, on stores, factories, hotels, warehouses, hospitals, public buildings, etc.

DESCRIPTION.

The exhaust steam on coming up out of the pipe strikes the cap, spreads out to the sides of the drum, is held in check by the flaring top, and falls in the shape of water on the bottom of the drum, and is carried off through the discharge pipe. The light vapory steam passes out at the top of the drum into the open air, making no impression on the surrounding objects.

The operation is absolutely effectual. It simply holds the

exhaust steam within the drum long enough to let the colder temperature of the atmosphere convert the heavy steam into water, and allow the light steam to pass away.

Much scalding hot water comes up the exhaust pipe with the waste steam, and is slushed out—quarts at a time—on the roof. This Trap takes all this water and conveys it off through the discharge pipe to the gutter, or to the leader pipe, and so saves the building.

ADVANTAGES.

It saves IRON FRONTS from one of the principal causes of rusting—prevents the wet steam from blowing over on the front. It saves a stone front from discoloring and ruin.

It saves the roof of the building by preventing the scalding water from falling and destroying the paint, and rusting the tin. It saves the surrounding brick walls and the chimneys.

It performs its work perfectly by condensing the heavy steam within the drum, and allows the light, vapory steam to pass away. Not a drop of water escapes. A constant and steady stream of water is caught in and carried off by the Trap when the engine is at work.

It cannot freeze up or get out of order. It will need no repairs, lasting as long as the material of which it is made will last.

It saves all annoyance to people passing by in the street, and at open windows, from the falling spray.

It permits no back pressure on the engine. The steam has a clear passage out.

It is compact, and sightly and durable. Its simplicity is one of its chief merits, and its effectiveness has been proved by its use for a number of years.

LIST OF SIZES OF EXHAUST TRAPS FOR STEAM PIPES.

SIZES.	DIMENSIONS.	Discharge Pipe.	Price $ put up complete.	Or, for the Trap only, $
No. 1. For 1 in. Exhaust. Pipe.	Drum 12×14 in. Cap 6 in. dia., sets 3 in. above top of Exhaust Pipe, and 4 in. from top of cap to top of drum.	¾ in.	$	$
No. 1½. For 1½ in. "	Drum 14×16 Cap 7 in. dia., sets 3½ in. above top of Exhaust Pipe, and 4½ in. from top of cap to top of drum.	" " ¾ "	"	" $
No. 2. For 2 in. "	Drum 16×18 Cap 8 in. dia., sets 4 in. above top of Exhaust Pipe, and 5 in. from top of cap to top of drum.	" " 1 "	"	" $
No. 2½. For 2½ in. "	Drum 18×20 Cap 9 in. dia., sets 4½ in. above top of Exhaust Pipe, and 5½ in. from top of cap to top of drum.	" " 1 "	"	" $
No. 3. For 3 in. "	Drum 18×20 Cap 10 in. dia., sets 5 in. above top of Exhaust Pipe, and 6 in. from top of cap to top of drum.	" " 1 "	"	" $
No. 4. For 4 in. "	Drum 20×24 Cap 12 in. dia., sets 6 in. above top of Exhaust Pipe, and 6½ in. from top of cap to top of drum.	" " 1¼ "	"	" $
No. 5. For 5 in. "	Drum 22×26 Cap 14 in. dia., sets 7 in. above top of Exhaust Pipe, and 7 in. from top of cap to top of drum.	" " 1¼ "	"	" $

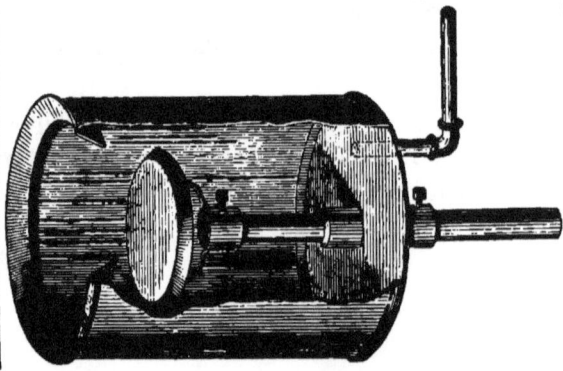

IRON PORT-HOLES FOR FIRE WALLS.

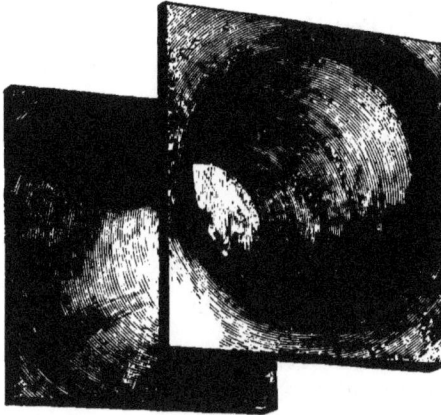

Extract from the New York Fire Law.

PARAPET WALL, to be at least five feet high above the roof, twelve inches thick and coped, and to have openings three and a half feet above the roof, suitable for fire defence.

If the opening is made too large, a fork of flame is liable to come through and fatally burn the fireman. If made small, it is inconvenient to see through and work in. The thickness of wall cuts off the side and downward angles of sight, and prevents the pipe-man from playing on the flames except at such angles as can be got through the oblong hole, which necessarily carries the stream of water not much nearer than the centre, and more often to the far side of the adjoining burning building.

This Port-Hole is made in one casting, in shape like a dice-box or an hour-glass, six inches diameter of opening in the centre, and radiating to a larger diameter of twelve inches on either side of the fire-wall. The small opening affords the fireman proper protection from the flames, and the large opening gives him increased angles of sight, free room to work in, and enables him to turn his stream of water in any direction. The work of battling with a fire is done more effectually.

In the small opening is placed a pane of mica, to enable the fireman to view the fire with safety from the flames and heat, in advance of playing on the burning mass. This mica pane fits in a small groove in the casting, and is held in place simply by a little putty. When the fireman puts through his pipe he at once knocks away the obstruction, and the full and clear diameter of the Port-Hole is at his service. The replacement of a new pane of mica is cheaply and quickly done. No sharp angles or corners on the iron, so when occasion requires, the hose may be easily dragged through, and without hindrance to the free flow of water.

Port-Holes should be placed about ten feet apart in the wall.

IRON.

Cast Iron expands $\frac{1}{162000}$ of its length for one degree of heat; greatest change in the shade in this climate, $\frac{1}{1170}$ of its length; exposed to the sun's rays, $\frac{1}{1000}$; shrinks in cooling from $\frac{1}{85}$ to $\frac{1}{98}$ of its length; is crushed by a force of 93,000 lbs. upon a square inch; will bear without permanent alteration, 15,300 lbs. upon a square inch, and an extension of $\frac{1}{1200}$ of its length.

Wrought Iron expands $\frac{1}{143000}$ of its length for one degree of heat; will bear on a square inch, without permanent alteration, 17,800 lbs., and an extension in length of $\frac{1}{14000}$; cohesive force is diminished $\frac{1}{80000}$ by an increase of one degree of heat.

Shrinkage of Castings.

The pattern-maker's rule should be for

Cast Iron	$\frac{1}{8}$ of an inch	longer per lineal foot.
Brass	$\frac{3}{16}$ " "	
Lead	$\frac{1}{4}$ " "	
Tin	$\frac{1}{12}$ " "	
Zinc	$\frac{5}{16}$ " "	

Iron, small cylinders.... = $\frac{1}{16}$ in. per ft.	Iron, in length.... = $\frac{1}{8}$ in 16 ins.
" Pipes .... = $\frac{1}{8}$ "	Brass, thin......... = $\frac{1}{8}$ in 9 ins.
" Girders, beams, etc. = $\frac{1}{8}$ in. 15 ins.	Brass, thick........ = $\frac{1}{8}$ in 10 ins.
" Large cylinders, the contraction of diam. at top. = $\frac{1}{16}$ per foot.	Zinc .... = $\frac{5}{16}$ in a foot.
	Lead............. = $\frac{5}{16}$ in a foot.
	Copper........... = $\frac{3}{16}$ in a foot.
" Ditto at bottom... = $\frac{1}{12}$ per foot.	Bismuth.......... = $\frac{5}{32}$ in a foot.

To Find the Weight of Castings from that of their Patterns.

Multiply weight of white pine pattern by 16 for cast iron.
Ditto   "   "   "   18 for brass.
Ditto   "   "   "   19 for copper
Ditto   "   "   "   25 for lead.

To Compute the Weight of Cast Iron, Wrought Iron, etc.

Cubic inches multiplied by	.263	=	lbs. av.	cast iron.
"	"	"	.281	=	"	wrought do.
"	"	"	.283	=	"	steel.
"	"	"	.3225	=	"	copper.
"	"	"	.3037	=	"	brass.
"	"	"	.26	=	"	zinc.
"	"	"	.4103	=	"	lead.
"	"	"	.2636	=	"	tin.
"	"	"	.4908	=	"	mercury.
Cylindrical in.	"	"	.2065	=	"	cast iron.
"	"	"	.2168	=	"	wrought iron.
"	"	"	.2223	=	"	steel.
"	"	"	.2533	=	"	copper.
"	"	"	.2385	=	"	brass.
"	"	"	.2042	=	"	zinc.
"	"	"	.3223	=	"	lead.
"	"	"	.207	=	"	tin.
"	"	"	.3854	=	"	mercury.
Avoirdupois lbs.	"	"	.000	=	cwts.
"	"	"	.00045	=	tons.

Cast Iron.—Example, for Plates, Etc.

What will a plate 12″ × 12″ × 1″ weigh ? Rule.—Ascertain the number of cubic inches in the piece, multiply them by .263 (the weight of a cubic inch as given in the above table) and the product will give the weight in pounds.

Thus: 12″ × 12″ × 1″ = 144 cubic inches.
.263
_____

37.872, say 38 lbs.

A short method : Rule.—Divide the number of cubic inches in the piece by 4, and to the product add a 20th.

Thus: $12 \times 12 \times 1 =$ 144
          divide 4)   ‾‾‾‾‾‾

                        36
          add $\frac{1}{20}$ 2   say
                      ‾‾‾‾‾
                        38   lbs.

EXAMPLE, FOR BOX COLUMNS, ETC.

What will a box column $12'' \times 12'' \times 1$ inch thick weigh
per lineal foot? Rule.—Ascertain the number of cubic inches
to the foot, multiply them by .263 and the product will give
the weight in pounds.

Thus: 12
        12
        12
        12
        ‾‾‾‾
        $48 \times 12''$ long $= 576$ cubic inches in the foot.
                        .263
                      ‾‾‾‾‾‾‾
                151.488 say 151 lbs.

A short method: Rule.—Multiply the number of cubic
inches in the area by 3, and to the product add a 20th.

Thus: 12
        12
        12
        12
        ‾‾‾‾
        $48 \times 1 =$ 48 cubic inches area.
                          3
                        ‾‾‾‾
                        144
          add $\frac{1}{20}$ 7
                        ‾‾‾‾
                        151 lbs.

EXAMPLE, FOR ROUND COLUMNS.

What will a round column 12 in. dia. and 1 in. thick weigh per foot? Rule.—Ascertain the number of cubic inches to the foot, multiply them by .263 and the product will give the weight in pounds. [The decimal .263 is used for convenience sake. The correct decimal, however, for cylindrical inches is .2065.]

Thus: 12 in. dia. = 37.69 in. circumference × 1 in. thick = 37.69 × 12 inches long = 452.28 cubic inches in the ft.

$$\begin{array}{r} .263 \\ \hline 118.9 \end{array}$$ say 119 lbs.

A short method: Rule.—Multiply the diameter by $3\frac{1}{7}$ to get the circumference.—Multiply the number of cubic inches in the area by 3, and add a 20th.

Thus: 12 in. dia.

$$\begin{array}{r} 3\frac{1}{7} \\ \hline 38 \\ 3 \\ \hline 114 \\ \text{add } \frac{1}{20} \quad 5 \\ \hline 119 \text{ lbs.} \end{array}$$

An approximate method: Multiply the diameter by 9.

EXAMPLE, FOR CAST IRON T BEAMS.

What will be the weight of a beam whose bottom flange is 12″ × 1½″, centre web 18″ × 1″, and top flange 3″ × 1″?

Get the number of cubic inches in a foot and multiply by .263.

Thus: 12″ × 1½″ = 18
16″ × 1″ = 16
3″ × 1″ = 3

37 × 12″ long = 444 cubic in. to the ft.

.263

say 117 lbs.

13

Short method, thus:

$$12 \times 1\tfrac{1}{2} = 18$$
$$16 \times 1 = 16$$
$$3 \times 1 = 3$$
$$\overline{37}$$
Multiply by $\quad3$
$$\overline{111}$$
add $\frac{1}{20}$ $\quad6$
$$\overline{117 \text{ lbs.}}$$

Wrought Iron.—Computing Weights.

The decimal for wrought iron is .281 in multiplying the number of cubic inches. The manner the same as given for cast iron.

For short methods of figuring $\frac{1}{10}$th is to be added to the product, instead of $\frac{1}{20}$, as in cast iron.

Thus: What will a plate of wrought iron 12″ × 12″ × 1 inch thick weigh?

	Short method.
12″ × 12″ × 1″ = 144	12″ × 12″ × 1″ = 144
.281	divide 4)
	36
40 lbs.	add $\frac{1}{10}$ 4
	40 lbs.

To Test the Quality of Bar Iron.

If fracture gives long silky fibres of leaden-gray hue, fibres cohering and twisting together before breaking, it may be considered a *tough soft iron*. A medium, even grain, mixed with fibres, a good sign. A short, blackish fibre indicates badly refined iron. A very fine grain denotes a *hard steely iron*, apt to be cold-short, hard to work with a file. Coarse grain, with brilliant crystalized fracture, yellow or brown spots, denotes a *brittle iron*, cold short, working easily when heated; welds easily. Cracks on the edge of bars, sign of *hot-short iron.*

The foreign substances which iron contains modify its essential properties. *Carbon* adds to its hardness, but destroys some of its qualities, and produces Cast Iron or Steel according to the proportion it contains. *Sulphur* renders it fusible, difficult to weld, and brittle when heated or "*hot short.*" *Phosphorus* renders it "*cold short.*"

WEIGHT OF A LINEAL FOOT OF ROUND AND SQUARE BAR IRON, IN POUNDS.

ROUND IRON.				SQUARE IRON.			
Inch.	lbs.	Inch.	lbs.	Inch.	lbs.	Inch.	lbs.
¼	.163	4⅞	63.1	¼	.208	4¼	60.8
⅜	.368	5	66.7	⅜	.468	4⅜	64.5
½	.654	5⅛	69.7	½	.833	4½	68.2
⅝	1.02	5¼	73.2	⅝	1.30,	4⅝	72.0
¾	1.47	5⅜	76.7	¾	1.87	4¾	75.9
⅞	2.00	5½	80.3	⅞	2.55	4⅞	80.0
1	2.61	5⅝	84.0	1	3.33	5	84.2
1⅛	3.31	5¾	87.8	1⅛	4.21	5⅛	88.5
1¼	4.09	5⅞	91.6	1¼	5.20	5¼	92.8
1⅜	4.94	6	95.6	1⅜	6.30		97.3
1½	5.89	6¼	103.7	1½	7.50	5½	101.9
1⅝	6.91	6½	112.2	1⅝	8.80	5⅝	106.6
1⅞	8.01	6¾	120.9	1¾	10.2	5¾	111.4
1	9.20	7	130.0	1⅞	11.7	5⅞	116.3
2	10.4	7¼	139 5	2	13.3	6	121.2
2⅛	11.8	7½	149.3	2⅛	15.0	6¼	131.6
2¼	13.2	7¾	159 5	2¼	16.8	6½	142.3
2⅜	14.7	8	169.9	2⅜	18.8	6¾	153.5
2½	16.3	8¼	180.7	2½	20.8	7	165.0
2⅝	18.0	8½	191.8	2⅝	22.9	7½	189.5
2¾	19.7	8¾	203.3	2¾	25.2	8	215.6
2⅞	21.6	9	215.0	2⅞	27.5	8½	243.4
3	23.5	9¼	227.2	3	30.0	9	272.8
3⅛	25.5	9½	239.6	3⅛	32.5	9½	303.9
3¼	27.6	9¾	252.4	3¼	35.2	10	336.8
3⅜	29.8	10	266.3	3⅜	37.9	10½	371.3
3½	32.0	10¼	278.9	3½	40.8	11	407.5
3⅝	34.4	10½	292.7	3⅝	43.8	11½	445.4
3¾	36.8	10¾	306.8	3¾	46.8	12	485.0
4	42.5	11	321.2	4	53.9		
4⅛	45.2	11¼	336.0	4⅛	57.3		
4¼	47.9	11½	351.1				
4⅜	50.8	11¾	366.5				
4½	53.7	12	382.2				
4⅝	56 8						
4¾	59 9						

WEIGHT OF A LINEAL FOOT OF FLAT BAR IRON, IN POUNDS.

NO. I.

Breadth in inches	■ Thickness in Fractions of inches. ■								
	¼	5/16	⅜	7/16	½	⅝	¾	⅞	1
1	.83	1.04	1.25	1.46	1.67	2.08	2.50	2.92	3.34
1⅛	.93	1.17	1.40	1.64	1.87	2.34	2.81	2.28	3.75
1¼	1.04	1.30	1.56	1.82	2.08	2.60	3.13	3.65	4.17
1⅜	1.14	1.43	1.72	2.00	2.29	2.87	3.44	4.01	4.59
1½	1.25	1.56	1.87	2.19	2.50	3.13	3.75	4.38	5.00
1⅝	1.35	1.69	2.03	2.37	2.71	3.39	4.07	4.70	5.43
1¾	1.46	1.82	2.19	2.55	2.92	3.65	4.38	5.11	5.84
1⅞	1.56	1.95	2.34	2.74	3.13	3.91	4.69	5.47	6.26
2	1.67	2.08	2.50	2.92	3.34	4.17	5.01	5.86	6.68
2⅛	1.77	2.21	2.66	3.10	3.55	4.43	5.32	6.21	7.10
2¼	1.87	2.34	2.81	3.28	3.76	4.69	5.63	6.57	7 52
2⅜	1.98	2.47	2.97	3.47	3.96	4·95	5·95	6.94	7 93
2½	2.08	2.60	3.13	3.65	4.17	5.21	6.26	7.30	8.35
2⅝	2.19	2.74	3.28	3.83	4.38	5.47	6.57	7.67	8.77
2¾	2.29	2.87	3.44	4.01	4.59	5.74	6.88	8.03	9.18
2⅞	2.40	3.00	3.60	4.20	4.80	6.00	7.20	8.40	9.60
3	2.50	3.13	3.75	4.38	5.01	6.26	7.51	8.76	10.02
3¼	2.71	3.39	4.07	4.74	5.43	6.78	8.14	9. 9	10.86
3½	2.92	3.65	4.38	5.11	5.84	7.30	8.76	10.23	11.69
3¾	3.13	3.91	4.68	5.47	6.26	7.82	9.39	10.95	12.52
4	3.34	4.17	5.00	5.84	6.68	8.35	10.02	11.69	13.36
4¼	3.54	4.43	5.32	6.21	7.09	8.87	10.64	12.42	14.19
4½	3.75	4·69	5.63	6.57	7.51	9.39	11.27	13.15	15.03
4¾	3.96	4·95	5.94	6.94	7.93	9.91	11.89	13.88	15.86
5	4.17	5.21	6.26	7.30	8.35	10.44	12.52	14.61	16.70
5¼	4.38	5.47	6.57	7.67	8.76	10.96	13.14	15.34	17.53
5½	4.59	5.73	6.88	8.03	9.18	11.48	13.77	16.07	18.37
5¾	4.80	6.00	7.20	8.40	9.60	12.00	14.40	16.80	19.20
6	5.01	6.25	7.51	8.76	10.02	12.53	15.03	17.53	20.05

Table No. II., on the following page, gives a different arrangement, which may be found more convenient for reference to get the weight of a lineal foot of flat bar iron, in pounds.

WEIGHT OF A LINEAL FOOT OF FLAT BAR IRON IN POUNDS.

NO. II.

Th'k.	Wid.	1 ft.	Th'k.	Wid.	1 ft.	Th'k.	Wid.	1 ft.	Th'k.	Wid.	1 ft.
inch.	inch.	lbs.	inch.	inch.	lbs.	inch.	inch.	lbs.	inch.	inch.	lbs.
1/4	1	0.8	3/8	2 3/4	3.5	1/2	4 1/4	7.2	5/8	6	12.7
1/4	1 1/4	1.1	3/8	3	3.8	1/2	4 1/2	7.6	3/4	1	2.5
1/4	1 1/2	1.3	3/8	3 1/4	4.1	1/2	4 3/4	8.0	3/4	1 1/4	3.2
1/4	1 3/4	1.5	3/8	3 1/2	4.4	1/2	5	8.4	3/4	1 1/2	3.8
1/4	2	1.7	3/8	3 3/4	4.8	1/2	5 1/4	8.9	3/4	1 3/4	4.4
1/4	2 1/4	1.9	3/8	4	5.1	1/2	5 1/2	9.3	3/4	2	5.1
1/4	2 1/2	2.1	3/8	4 1/4	5.4	1/2	5 3/4	9.7	3/4	2 1/4	5.7
1/4	2 3/4	2.3	3/8	4 1/2	5.7	1/2	6	10.1	3/4	2 1/2	6.3
1/4	3	2.5	3/8	4 3/4	6.0	5/8	1	2.1	3/4	2 3/4	7.0
1/4	3 1/4	2.7	3/8	5	6.3	5/8	1 1/4	2.6	3/4	3	7.6
1/4	3 1/2	3.0	3/8	5 1/4	6.7	5/8	1 1/2	3.2	3/4	3 1/4	8.2
1/4	3 3/4	3.2	3/8	5 1/2	7.0	5/8	1 3/4	3.7	3/4	3 1/2	8.9
1/4	4	3.4	3/8	5 3/4	7.3	5/8	2	4.2	3/4	3 3/4	9.5
1/4	4 1/4	3.6	3/8	6	7.6	5/8	2 1/4	4.8	3/4	4	10.1
1/4	4 1/2	3.8	1/2	1	1.7	5/8	2 1/2	5.3	3/4	4 1/4	10.8
1/4	4 3/4	4.0	1/2	1 1/4	2.1	5/8	2 3/4	5.8	3/4	4 1/2	11.4
1/4	5	4.2	1/2	1 1/2	2.5	5/8	3	6.3	3/4	4 3/4	12.0
1/4	5 1/4	4.4	1/2	1 3/4	3.0	5/8	3 1/4	6.9	3/4	5	12.7
1/4	5 1/2	4.6	1/2	2	3.4	5/8	3 1/2	7.4	3/4	5 1/4	13.3
1/4	5 3/4	4.9	1/2	2 1/4	3.8	5/8	3 3/4	7.9	3/4	5 1/2	13.9
1/4	6	5.1	1/2	2 1/2	4.2	5/8	4	8.4	3/4	5 3/4	14.6
3/8	1	1.3	1/2	2 3/4	4.6	5/8	4 1/4	9.0	3/4	6	15.2
3/8	1 1/4	1.6	1/2	3	5.1	5/8	4 1/2	9.5	1	1 1/2	5.1
3/8	1 1/2	1.9	1/2	3 1/4	5.5	5/8	4 3/4	10.0	1	2	6.8
3/8	1 3/4	2.2	1/2	3 1/2	5.9	5/8	5	10.6	1	3	10.1
3/8	2	2.5	1/2	3 3/4	6.3	5/8	5 1/4	11.1	1	4	13.5
3/8	2 1/4	2.9	1/2	4	6.8	5/8	5 1/2	11.6	1	5	16.9
3/8	2 1/2	3.2				5/8	5 3/4	12.1	1	6	20.3

Table No. I., on the preceeding page, gives a different arrangement of weights of a lineal foot of flat bar iron, in pounds.

TO CALCULATE VALUE PER TON OF 2,240 POUNDS,

AT 1/16 OF A CENT PER POUND TO 13 CENTS PER POUND.

	$	c.		$	c.		$	c.		$	c.		$	c.		$	c.		$	c.
1/16	1	40	1 7/8	42	00	3 3/4	84	00	5 5/8	126	00	7 1/2	168	00	9 3/8	210	00	11 1/4	252	0.
1/8	2	80	2	44	80	3 7/8	86	00	5 3/4	128	80	7 5/8	170	80	9 1/2	212	80	11 3/8	254	80
1/4	5	60	2 1/8	47	60	4	89	60	5 7/8	131	60	7 3/4	173	60	9 5/8	215	60	11 1/2	257	60
3/8	8	40	2 1/4	50	40	4 1/8	92	40	6	134	40	7 7/8	176	40	9 3/4	218	40	11 5/8	260	40
1/2	11	20	2 3/8	53	20	4 1/4	95	20	6 1/8	137	20	8	179	20	9 7/8	221	20	11 3/4	263	20
5/8	14	00	2 1/2	56	00	4 3/8	98	00	6 1/4	140	00	8 1/8	182	00	10	224	00	11 7/8	266	00
3/4	16	80	2 5/8	58	80	4 1/2	100	80	6 3/8	142	80	8 1/4	184	80	10 1/8	226	80	12	268	80
7/8	19	60	2 3/4	61	60	4 5/8	103	60	6 1/2	145	60	8 3/8	187	60	10 1/4	229	60	12 1/8	271	60
1	22	40	2 7/8	64	40	4 3/4	106	40	6 5/8	148	40	8 1/2	190	40	10 3/8	232	40	12 1/4	274	40
1 1/8	25	20	3	67	20	4 7/8	109	20	6 3/4	151	20	8 5/8	193	20	10 1/2	235	20	12 3/8	277	20
1 1/4	28	00	3 1/8	70	00	5	112	00	6 7/8	154	00	8 3/4	196	00	10 5/8	238	00	12 1/2	280	00
1 3/8	30	80	3 1/4	72	80	5 1/8	114	80	7	156	80	8 7/8	198	80	10 3/4	240	80	12 5/8	282	80
1 1/2	33	60	3 3/8	75	60	5 1/4	117	60	7 1/8	159	60	9	201	60	10 7/8	243	00	12 3/4	285	60
1 5/8	36	40	3 1/2	78	40	5 3/8	120	40	7 1/4	162	40	9 1/8	204	40	11	246	40	12 7/8	288	40
1 3/4	39	20	3 5/8	81	20	5 1/2	123	20	7 3/8	165	20	9 1/4	207	29	11 1/8	249	20	13	291	20

NUMBER OF FEET IN A BUNDLE.

HOOP IRON. 1/2 cwt. (56 lb.) bundles.	
In. No.	Feet
5/8 × 21	817
3/4 × 20	633
7/8 × 19	451
1 × 18	361
1 1/8 × 17	275
1 1/4 × 16	215
1 1/2 × 15	159
1 3/4 × 15	137
2 × 14	108

SCROLL IRON.	
1/2 × 10	239
5/8 × 16	430
5/8 × 14	345
5/8 × 10	191
3/4 × 16	359
3/4 × 14	287
3/4 × 12	205
3/4 × 10	159
7/8 × 16	311
7/8 × 14	247
7/8 × 12	177
7/8 × 10	138
1 × 16	269
1 × 14	215
1 × 12	153
1 × 10	120

BAND IRON. 1 cwt. (112 lb.) bundles.			
In. No.	Feet	In. No.	Feet
1 1/8 × 12	264	2 3/4 × 12	111
1 1/8 × 10	212	2 3/4 × 10	87
1 1/8 × 7	159	2 3/4 × 8	73
1 1/4 × 12	245	2 3/4 × 6	60
1 1/4 × 10	191	3 × 12	102
1 1/4 × 7	143	3 × 10	79
1 1/2 × 12	204	3 × 8	65
1 1/2 × 10	159	3 × 6	55
1 1/2 × 7	119	3 1/4 × 10	74
1 3/4 × 12	176	3 1/4 × 8	60
1 3/4 × 10	137	3 1/4 × 6	51
1 3/4 × 8	112	3 1/2 × 10	68
1 3/4 × 7	102	3 1/2 × 8	56
2 × 12	153	3 1/2 × 6	47
2 × 10	119	4 × 10	59
2 × 8	98	4 × 8	49
2 × 7	89	4 × 6	41
2 × 6	82	4 1/2 × 10	53
2 1/4 × 12	136	4 1/2 × 8	43
2 1/4 × 10	106	4 1/2 × 6	36
2 1/4 × 8	87	5 × 10	47
2 1/4 × 6	73	5 × 8	39
2 1/2 × 12	122	5 × 6	33
2 1/2 × 10	96	6 × 10	39
2 1/2 × 8	78	6 × 8	33
2 1/2 × 6	66	6 × 6	27

FLAT IRON. 1 cwt. (112 lb.) bundles.		
In.		Feet
1/2	1/4	268
1/2	5/16	215
1/2	3/8	176
5/8	1/4	215
5/8	5/16	172
5/8	3/8	143
5/8	1/2	108
3/4	1/4	176
3/4	5/16	143
3/4	3/8	119
3/4	7/16	102
3/4	1/2	90
3/4	5/8	71
7/8	1/4	154
7/8	5/16	123
7/8	3/8	102
7/8	7/16	88
7/8	1/2	77
7/8	5/8	61
1	1/4	134
1	5/16	107
1	3/8	89
1	7/16	77
1	1/2	67
1	9/16	59
1	5/8	53

ROUND IRON. 1 cwt. (112 lb.) bundles.	
Inch.	Feet.
3/16	1116
1/4	687
5/16	439
3/8	304
7/16	223
1/2	171
9/16	135
5/8	109
11/16	91
3/4	76

SQUARE IRON.	
3/16	956
1/4	538
5/16	344
3/8	239
7/16	175
1/2	134
9/16	106
5/8	86
11/16	71
3/4	69

NOTE.—This table is calculated for exact size. Rolled Iron is usually full size, for which allowance should be made.

SHEET IRON.

WEIGHT OF A SUPERFICIAL FOOT.

B. W. Gauge.	Dec. of an inch.	Weight in lbs.	B. W. Gauge.	Dec. of an inch.	Weight in lbs.
00000 ($\frac{1}{2}$)	.500	20.208	16 ($\frac{1}{16}$)	.063	2.546
0000	.450	18.187	17	.055	2.223
000	.437	17.662	18	.048	1.940
00 ($\frac{3}{8}$)	.375	15.156	19	.042	1.697
0	.340	13.742	20	.035	1.415
1 ($\frac{5}{16}$)	.312	12.610	21	.033	1.334
2	.284	11.477	22	.029	1.172
3	.261	10.549	23	.028	1.132
3—4 ($\frac{1}{4}$)	.250	10.104	24	.025	1.014
4	.239	9.660	25	.021	.849
5	.217	8.770	26	.020	.808
6	.208	8.407	27	.018	.727
7 ($\frac{3}{16}$)	.187	7.558	28	.015	.606
8	.166	6.709	29	.013	.525
9	.158	6.386	30	.012	.485
10	.137	5.537	31	.010	.404
11 ($\frac{1}{8}$)	.125	5.052	32	.009	.364
12	.109	4.405	33	.008	.323
13	.094	3.799	34	.007	.283
14	.080	3.233	35	.005	.202
15	.072	2.910	36	.004	.162

HOOP IRON.

DIMENSIONS AND WEIGHT IN LBS. PER FOOT RUN.

Breadth	$\frac{5}{8}$	$\frac{3}{4}$	$\frac{7}{8}$	1 in.	1$\frac{1}{8}$	1$\frac{1}{4}$
B. W. Gauge	21	20	19	18	17	16
Weight per lineal foot	.0666	.0875	.1216	.1636	.21	.27

Breadth	1$\frac{3}{8}$	1$\frac{1}{2}$	1$\frac{3}{4}$	2 in.	2$\frac{1}{4}$	2$\frac{1}{2}$
B. W. Gauge	15	15	14	13	13	12
Weight per lineal foot	33	.36	.484	.634	.714	.91

WEIGHT OF BOILER IRON.

$\frac{1}{8}$ inch iron weighs	5 pounds per square foot.
$\frac{3}{16}$ " " "	7$\frac{1}{2}$ " " " "
$\frac{1}{4}$ " " "	10 " " " "
$\frac{5}{16}$ " " "	12$\frac{1}{2}$ " " " "
$\frac{3}{8}$ " " "	15 " " " "
$\frac{7}{16}$ " " "	17$\frac{1}{2}$ " " " "
$\frac{1}{2}$ " " "	20 " " " "

WEIGHT OF ANGLE IRON.

PER LINEAL FOOT.

	Lbs.		Lbs.
$\frac{3}{4} \times \frac{3}{4} \times \frac{1}{8}$	.63	$2\frac{1}{2} \times 2\frac{1}{2} \times \frac{5}{16}$	5.
$1 \times 1 \times \frac{1}{8}$	1.	$3 \times 3 \times \frac{3}{8}$	7.
$1\frac{1}{4} \times 1\frac{1}{4} \times \frac{3}{16}$	1.50	$3\frac{1}{2} \times 3\frac{1}{2} \times \frac{7}{16}$	9.
$1\frac{1}{2} \times \frac{1}{2} \times 1\frac{3}{16}$	2.	$4 \times 4 \times \frac{1}{2}$	12.50
$1\frac{3}{4} \times 1\frac{3}{4} \times \frac{3}{16}$	2.75	$6 \times 4 \times \frac{1}{2}$	17.
$2 \times 2 \times \frac{1}{4}$	3.50	$6 \times 6 \times \frac{1}{2}$	20.
$2\frac{1}{4} \times 2\frac{1}{4} \times \frac{1}{4}$	4.25		

WEIGHT OF TEE IRON.

PER LINEAL FOOT.

	Lbs.		Lbs.
$\frac{3}{4} \times \frac{3}{4} \times \frac{1}{8}$	.63	$2\frac{1}{4} \times 2\frac{1}{4} \times \frac{1}{4}$	4.
$1 \times 1 \times \frac{1}{8}$	.95	$2\frac{1}{2} \times 2\frac{1}{2} \times \frac{5}{16}$	4.87
$1\frac{1}{4} \times 1\frac{1}{4} \times \frac{1}{4}$	2.25	$3 \times 3 \times \frac{3}{8}$	7.50
$1\frac{1}{2} \times 1\frac{1}{2} \times \frac{1}{4}$	2.63	$3\frac{1}{2} \times 3\frac{1}{2} \times \frac{3}{8}$	9.50
$1\frac{3}{4} \times 1\frac{3}{4} \times \frac{1}{4}$	3.08	$3\frac{1}{2} \times 3\frac{1}{2} \times \frac{1}{2}$	10.50
$2 \times 2 \times \frac{1}{4}$	3.40	$4 \times 4 \times \frac{1}{2}$	14.

GALVANIZED AND BLACK IRON.

WEIGHT IN POUNDS PER SQUARE FOOT OF GALVANIZED SHEET-IRON, BOTH FLAT AND CORRUGATED.

The numbers and thicknesses are those of the iron before it is galvanized. When a flat sheet (the ordinary size of which is from 2 to $2\frac{1}{2}$ feet in width, by 6 to 8 feet in length) is converted into a corrugated one, with corrugations 5 inches wide from centre to centre, and about an inch deep (the common sizes), its width is thereby reduced about $\frac{1}{10}$th part, or from 30 to 27 inches; and consequently the weight per square foot of area covered is increased about $\frac{1}{9}$th part. When the corrugated sheets are laid upon a roof, the overlapping of about $2\frac{1}{2}$ inches along their sides, and of four inches along their ends, diminishes the covered area about $\frac{1}{7}$th part more; making their weight per square foot of roof about $\frac{1}{6}$th part greater than before. Or the weight of corrugated iron per square foot in place on a roof, is about $\frac{1}{3}$ greater than that of the flat sheets of above sizes of which it is made.

Number by Birmingham Wire Gauge.	BLACK.		GALVANIZED.		
	Thickness in Inches.	Flat. Lbs.	Flat. Lbs.	Corrugated. Lbs.	Cor. on Roof. Lbs.
30	.012	.485	.806	.896	1.08
29	.013	.526	.857	.952	1.14
28	.014	.565	.897	.997	1.20
27	.016	.646	.978	1.09	1.30
26	.018	.722	1.06	1.18	1.41
25	.020	.808	1.14	1 27	1.52
24	.022	.889	1.22	1.36	1.62
23	.025	1.01	1.34	1.49	1.79
22	.028	1.13	1.46	1.62	1.95
21	.032	1.29	1.63	1.81	2.17
20	.035	1.41	1.75	1.94	2.33
19	.042	1.69	2.03	2 26	2.71
18	.049	1.98	2.32	2.58	3.09
17	.058	2.34	2.68	2.98	3.57
16	.065	2.63	2.96	3.29	3.95
15	.072	2.91	3.25	3.61	4.33
14	.083	3.36	3.69	4.10	4.92
13	.095	3.84	4.18	4.64	5.57

Note.—The galvanizing of sheet-iron adds about one-third of a pound to its weight per square foot.

CORRUGATED IRON ROOFING.

Birmingham Wire Gauge.	Size of Sheets.	Weight per Square.	Number of Superficial Ft. per Ton.
	ft. ft. ft. ft.	cwts.	
16	6 × 2 to 8 × 3	$3\frac{1}{8}$	800
18	6 × 2 to 8 × 3	$2\frac{1}{4}$	1000
20	6 × 2 to 8 × 3	$1\frac{3}{4}$	1250
22	6 × 2 to 7 × $2\frac{1}{2}$	$1\frac{1}{2}$	1550
24	6 × 2 to 7 × $2\frac{1}{2}$	$1\frac{1}{4}$	1880
26	6 × 2 to 7 × $2\frac{1}{2}$	1	2170

IRON RIVETS.
Weight per 100.

Length Under Head.	DIAMETERS.						
	¼	⅜	½	⅝	¾	⅞	1
1	1.895	4.848	9.66	16.79	26.49	39.3	55.2
⅛	2.067	5.235	10.34	17.86	27.99	41.4	57.9
¼	2.238	5.616	11.04	18.96	29.61	43.5	60.7
⅜	2.410	6.003	11.73	20.03	31.13	45.6	63.4
½	2.582	6.402	12.43	21.04	32.74	47.8	66.2
⅝	2.754	6.789	13.12	22.11	34.25	49.9	68.9
¾	2.926	7.179	13.81	23.21	35.86	52.0	71.7
⅞	3.098	7.566	14.50	24.28	37.37	54.1	74.4
2	3.269	7.956	15.19	25.48	38.99	56.3	77.2
⅛	3.441	8.343	15.88	26.56	40.40	58.4	79.9
¼	3.613	8.733	16.57	27.65	42.11	60.5	82.7
⅜	3.785	9.120	17.26	28.73	43.67	62.6	85.4
½	3.957	9.511	17.95	29.82	45.24	64.8	88.2
⅝	4.129	9.898	18.64	30.90	46.80	66.9	90.9
¾	4.301	10.29	19.33	31.99	48.36	69.0	93.7
⅞	4.473	10.67	20.02	33.08	49.92	71.1	96.4
3	4.644	11.06	20.71	34.18	51.49	73.3	99.2
⅛	4.816	11.44	21.40	35.27	53.05	75.4	101.9
¼	4.988	11.84	22.09	36.35	54.61	77.5	104.7
⅜	5.160	12.23	22.78	37.44	56.17	79.6	107.4
½	5.332	12.62	23.48	38.52	57.74	81.8	110.2
⅝	5.504	13.01	24.17	39.60	59.30	83.9	112.9
¾	5.676	13.39	24.86	40.69	60.86	86.0	116.7
⅞	5.848	13.78	25.55	41.78	62.42	88.1	119.4
4	6.019	14.17	26.24	42.87	63.99	90.3	121.2
⅛	6.191	14.56	26.93	43.94	65.55	92.4	123.9
¼	6.363	14.95	27.62	45.01	67.11	94.5	126.6
100 Heads.	.519	1.74	4.14	8.10	13.99	22.27	33.15

Length of Rivet required to make one Head = 1½ diameters of Round Bar.

WEIGHTS AND MEASURES.

AVOIRDUPOIS WEIGHT.
16 drachms.............= 1 ounce.
16 ounces.............. = 1 pound.
28 pounds........... .. = 1 quarter.
4 quarters (112 lbs.)...... = 1 cwt.
20 cwt (2240 lbs.)........ = 1 ton.

CUBIC OR SOLID MEASURE.
1 cubic foot = 1728 cubic inches.
1 " " = 2200 cylindrical inches.
1 " " = 3300 spherical "
1 " yard = 27 cubic feet.

LONG MEASURE.
12 inches		1 foot						
36	=	3	=	1 yard				
72	=	6	=	2	=	1 fathom		
198	=	16.5	=	5.5	=	2.75	=	1 perch or pole
7920	=	660	=	220	=110		= 40	=1 furlong
63360	=5280		=1760	=880		=320	=8	=1 mile.

WEIGHT OF 100 BOLTS WITH SQUARE HEADS AND NUTS.

Length under Head.	DIAMETER OF BOLTS.							
	¼ in.	⅜ in.	½ in.	⅝ in.	¾ in.	⅞ in.	1 in.	1⅛ in.
	lbs.	lbs.	lbs.	lbs.	lbs.	lbs.	lbs.	lbs.
1 inch	3½	9	20	32	......	......	......	......
1¼ "	3⅞	9⅞	21	34½	......	......	......	......
1½ "	4¼	10¾	22	37	......	......	......	......
1¾ "	4⅝	11⅝	23	39½	......	......	......	......
2 "	5	12½	24	42	70	130	180	......
2¼ "	5⅜	13½	25½	44½	73½	132½	185	......
2½ "	5¾	14⅝	27	47	77	135	190	......
2¾ "	6⅛	15¼	28½	49½	80½	137½	195	......
3 "	6½	16¼	30	52	84	140	200	296
3½ "	7⅛	18⅛	33	56½	90	148	210	310
4 "	7¾	20	36	61	196	156	220	324
4½ "	8⅜	21⅝	39	65½	101½	164	230	338
5 "	9	23¼	42	70	107	172	240	352
5½ "	9¾	24⅞	45	74	112½	180	251	366
6 "	10⅜	26½	48	78	118	188	262	370
7 "	11¾	29½	54	86	130	204	284	384
8 "	13⅛	33	60	94	143	220	306	398
9 "	14½	36	66	102	156	236	328	426
10 "	16	40	72	110	170	252	350	454
11 "	17¼	43	78	118	185	268	372	482
12 "	18⅝	46	84	127	200	284	393	510
13 "	......	......	92	155	219	335	426	538
14 "	......	......	97	163	237	351	448	566
15 "	......	......	103	170	249	391	470	594

WEIGHTS OF NUTS AND BOLT-HEADS IN LBS.

FOR CALCULATING THE WEIGHT OF LONGER BOLTS.

Diameter of Bolt, in inches.	¼	⅜	½	⅝	¾	⅞	1	1¼	1½	1¾	2	2½	3
Weight of Hexagon Nut and Head..........	.017	.057	.128	.267	.43	.73	1.10	2.14	3.78	5.6	8.75	17.	29.
Weight of Square Nut and Head..	.021	.069	.164	.320	.55	.88	1.31	2.56	4.42	7.0	10.5	21.	36.

STANDARD SIZES OF WASHERS.

NUMBER IN 100 LBS.

Diameter.	Size of Hole.	Thickness Wire Gauge.	Size of Bolt.	Number in 100 lbs.
Inch.	Inch.	No.	Inch.	
5/8	5/16	16	1/4	29300
3/4	3/8	16	5/16	18000
1	7/16	14	3/8	7600
1½	9/16	11	1/2	3300
1½	5/8	11	9/16	2180
1½	11/16	11	5/8	2350
1¾	13/16	11	3/4	1680
2	31/32	10	7/8	1140
2½	1⅛	8	1	580
2¾	1¼	8	1⅛	470
3	1⅜	7	1¼	360
3	1½	6	1⅜	360

RELATIVE WEIGHTS OF METALS.

The weight of Bar Iron being 1,

Weight of Cast Iron	=	.95
" " Steel	=	1.02
" " Copper	=	1.16
" " Brass	=	1.09
" " Lead	=	1.48

The weight of Cast Iron being 1,

Weight of Bar Iron	=	1.07
" " Steel	=	1.08
" " Brass	=	1.16
" " Copper	=	1.21
" " Lead	=	1.56

VARIOUS METALS.

THE WEIGHT OF A SUPERFICIAL FOOT.

Thickness in inches.	Wrought Iron.	Cast Iron.	Steel.	Copper.	Brass.	Lead.	Zinc.	Thickness in inches.
	lbs.	lbs.	lbs.	lbs.	lbs.	lbs.	lbs.	
1/16	2.526	2.344	2 552	2.891	2.734	3.708	2.344	1/16
1/8	5.052	4.687	5.104	5.781	5.469	7.417	4.687	1/8
3/16	7.578	7.031	7.656	8.672	8.203	11.125	7.031	3/16
1/4	10.104	9.375	10.208	11.563	10.938	14.833	9.375	1/4
5/16	12.630	11.719	12.760	14.453	13.672	18.542	11.719	5/16
3/8	15.156	14.062	15.312	17.344	16.406	22.250	14.062	3/8
7/16	17.682	16.406	17.865	20.234	19.141	25.958	16.406	7/16
1/2	20.208	18.750	20.417	23.125	21.875	29.667	18.750	1/2
9/16	22.734	21.094	22.969	26.016	24.609	33.375	21.094	9/16
5/8	25.260	23.437	25.521	28.906	27.344	37.083	23.437	5/8
11/16	27.786	25.781	28.073	31.797	30.078	40.792	25.781	11/16
3/4	30.312	28.125	30.625	34.688	32.813	44.500	28 125	3/4
13/16	32.839	30.469	33.177	37.578	35.547	48.208	30.469	13/16
7/8	35.365	32.812	35.729	40.469	38.281	51.917	32.812	7/8
15/16	37.891	35.156	38.281	43.359	41.016	55.625	35.156	15/16
1	40.417	37.500	40.833	46.250	43.750	59.333	37.500	1

CAST IRON BALLS.

Inches— Diam.	Lbs. Weight.	Inches— Diam.	Lbs. Weight.	Inches— Diam.	Lbs. Weight.
½	.07	4½	12.43	8½	83.77
1	.14	5	17.05	9	99.44
1¼	.46	5½	22.70	9½	116.9
2	1.09	6	29.47	10	136.4
2½	2.13	6½	37 46	10½	157.9
3	3.68	7	46.80	11	181.6
3½	5.85	7½	57.57	11½	207.4
4	8.73	8	69.80	12	235.7

WEIGHT OF

SOLID CAST METAL CYLINDERS.

EACH 1 FOOT IN LENGTH.

Diameter.	Iron.	Copper.	Brass.	Lead.
Inches.	lbs.	lbs.	lbs.	lbs.
1	2.5	3.0	2.9	3.9
2	9.8	12.0	11.4	15.5
3	22.1	27.0	25.8	34.8
4	39.3	47.9	45.8	61 9
5	61.4	74.9	71.6	96.7
6	88.4	107.8	103.0	139.3
7	120.3	146.8	140.2	189.6
8	157.1	191.7	183.2	247.7
9	198.8	242.7	231.8	313.4
10	245.4	299.5	286.2	387.0

DIFFERENT COLORS OF IRON CAUSED BY HEAT.

C.	FAHR.	COLOR.
210°	....410°	....Pale Yellow.
221	430	....Dull Yellow.
256	493	....Crimson.

261 ....502 ⎫ Violet, Purple and dull Blue; between 261° C. to 370° C.
370 ....680 ⎬ ·· it passes to Bright Blue, to Sea Green, and then disap-
        ⎭ pears.

500 ....932 ....Commences to be covered with a light coating of oxide;
loses a good deal of its hardness, becomes a good deal
more impressible to the hammer and can be twisted with
ease.

525 .... 977 ....Becomes Nascent Red.
700 ....1292 ....Sombre Red.
800 ....1472 ....Nascent Cherry.
900 ....1657 ....Cherry.

C.	FAHR.	COLOR.
1000	1832	....Bright Cherry.
1100	2012	....Dull Orange.
1200	...2192	....Bright Orange.
1300	...2372	....White.
1400	...2552	....Brilliant White—welding heat.
1500	....2732 }	....Dazzling White.
1600	....2912 }	

MELTING POINT OF METALS.

NAME.	FAHR.	FAHR.	
Platina	..........4593°		
Antimony	 955	842	
Bismuth	 487	507	
Tin (average)	 475		
Lead "	 622	620	
Zinc	 772	782	
Cast Iron	2786	} 1922..2012 White.	
		} 2012..2192 Gray.	
Wrought Iron	2552	......2733 Welding heat.	
Copper (average)	...2174		

WEIGHTS OF MATERIALS.

	Per Cubic Foot.
Water	62.3
Fire-brick	137.
Brick-work	112.
Coal, Anthracite	100.
" Bituminous	77 to 90
Coke	62 to 104
Granite	164—172
Plaster of Paris	73.5
Limestone	169—175
Masonry	116—144
Sandstone	144
Slate	178
Common Gravel	109
Mud	102
Mortar	98
Concrete	125
Common Soil	137
Glass	165

THE RELATIVE CONDUCTING POWER OF MATE-RIALS USED IN BUILDING.

Slate	....................100.		Brick, common	.... 60.14
Plaster of Paris	.... 20.26		" fire	.... 61.70
Plaster and sand	.... 18.70		Bathstone	.... 61.08
Roman cement	.... 20.80		Oak	.... 33.66
Lath and plaster	.... 25.55		Fir	.... 27.60
Asphalte	.... 45.19		Beech	.... 22.44
Chalk	.... 56.38		Lead	....................521.34

CIRCUMFERENCES OF CIRCLES.

ADVANCING BY EIGHTHS.

CIRCUMFERENCES.

Diam	.0	.⅛	.¼	.⅜	.½	.⅝	.¾	.⅞
0	.0	.3927	.7854	1.178	1.570	1.963	2.356	2.748
1	3.141	3.534	3.927	4.319	4.712	5.105	5.497	5.890
2	6.283	6.675	7.068	7.461	7.854	8.246	8.639	9.032
3	9.424	9.817	10.21	10.60	10.99	11.38	11.78	12.17
4	12.56	12.95	13.35	13.74	14.13	14.52	14.92	15.31
5	15.70	16.10	16.49	16 88	17.27	17.67	18.06	18.45
6	18.84	19 24	19.63	20.02	20.42	20.81	21.20	21.59
7	21.99	22.38	22.77	23.16	23 56	23.95	24.34	24.74
8	25.13	25.52	25.91	26.31	26.70	27.09	27.48	27.88
9	28.27	28.66	29.05	29.45	29.84	30.23	30.63	31.02
10	31.41	31.80	32.20	32.59	32.98	33.37	33.77	34.16
11	34.55	34.95	35.34	35.73	36.12	36.52	36.91	37.30
12	37.69	38.09	38.48	38.87	39.27	39.66	40.05	40.44
13	40.84	41.23	41.62	42.01	42.41	42.80	43.19	43.58
14	43.98	44.37	44.76	45.16	45.55	45.94	46.33	46.73
15	47.12	47.51	47.90	48.30	48.69	49.08	49.48	49.87
16	50.26	50.65	51.05	51.44	51.83	52.22	52.62	53.01
17	53.40	53.79	54.19	54 58	54.97	55.37	55.76	56.15
18	56.54	56.94	57.33	57.72	58.11	58.51	58.90	59.29
19	59.69	60.08	60.47	60.86	61.26	61.65	62.04	62.43
20	62.83	63.22	63.61	64.01	64.40	64.79	65.18	65.58
21	65.97	66.36	66.75	67.15	67.54	67 93	68.32	98.72
22	69.11	69.50	69.90	70.29	70.68	71.07	71.47	71.86
23	72.25	72.64	73.04	73.43	73.82	74.22	74.61	75.00
24	75.39	75.79	76.18	76.57	76.96	77.36	77.75	78.14
25	78.54	78.93	79.32	79.71	80.10	80.50	80.89	81.28
26	81.68	82.07	82.46	82.85	83.25	83.64	84.03	84.43
27	84.82	85.21	85.60	86.00	86.39	86.78	87.17	87.57
28	87.96	88.35	88.75	89.14	89.53	89.92	90.32	90.71
29	91.10	91.49	91.89	92.28	92.67	93.06	93.46	93.85
30	94.24	94.94	95.03	95.42	95.81	96.21	96.60	96.99
31	97.39	97.78	98.17	98.57	98.96	99.35	99.75	100.14
32	100.53	100.92	101.32	101.71	102.10	102.49	102.89	103.29
33	103.67	104.07	104.46	104.85	105.24	105.64	106.03	106.42
34	106.81	107.21	107.60	107.99	108.39	108.78	109.17	109.56
35	109.96	110.35	110.74	111.13	111.53	111.92	112.31	112.71
36	113.10	113.49	113.88	114.28	114.67	115 06	115.45	115.85
37	116.24	116.63	117.02	117.42	117.81	118.20	118.60	118.99
38	119.38	119.77	120.17	120.56	120.95	121.34	121.74	122.13
39	122.52	122.92	123.31	123.70	124.09	124.49	124.88	125.27
40	125.66	125.06	126.45	126.84	127.24	127.63	128.02	128.41
41	128.81	129.20	129.59	129.98	130.38	130.77	131.16	131.55
42	131.95	132.34	132.73	133.13	133.52	133.91	134.30	134.70
43	135.00	135.48	135.87	136.27	136.66	137.05	137.45	137.84
44	138.23	138.62	139.02	139.41	139.80	140.19	140.59	140.98
45	141.37	141.76	142.16	142.55	142.94	143.34	143.73	144.12

ARCHITECTURAL IRON WORK.

AREAS OF CIRCLES.

ADVANCING BY EIGHTHS.

				AREAS.				
Diam	.0	.⅛	.¼	.⅜	.½	.⅝	.¾	.⅞
0	.0	.0122	.0490	.1104	.1963	.3008	.4417	.6013
1	.7854	.9940	1.227	1.484	1.767	2.073	2.405	2.761
2	3.1416	3.546	3.976	4.430	4.908	5.411	5.939	64.91
3	7.068	7.669	8.285	8.946	9.621	10.32	11.04	11.79
4	12.56	13.36	14.18	15.03	15.90	16.80	17.72	18.66
5	19.63	20.62	21.64	22.69	23.75	24.85	25.96	27.10
6	28.27	29.46	30.67	31.91	33.18	34.47	35.78	37.12
7	38.48	39.87	41.28	42.71	44.17	45.60	47.17	48.70
8	50.26	51.84	53.45	55.08	56.74	58.42	60.13	61.86
9	63.61	65.39	67.20	69.02	70.88	72.75	74.69	76.58
10	78.54	80.51	82.51	84.54	86.59	88.66	90.76	92.88
11	95.03	97.20	99.40	101.6	103.8	106.1	108.4	110.7
12	113.0	115.4	117.8	120.2	122.7	125.1	127.6	130.1
13	132.7	135.2	137.8	140.5	143.1	145.8	148.4	151.2
14	153.9	156.6	159.4	162.2	165.1	167.9	170.8	173.7
15	176.7	179.6	182.6	185.6	188.6	191.7	194.8	197.9
16	201.0	204.2	207.3	210.5	213.8	217.0	220.3	223.6
17	226.9	230.3	233.7	237.1	240.5	243.9	247.4	250.9
18	254.4	258.0	261.5	265.1	268.8	272.4	276.1	279.8
19	283.5	287.2	291.0	294.8	298.6	302.4	306.3	310.2
20	314.1	318.1	322.0	326.0	330.0	334.1	338.1	342.2
21	346.3	350.4	354.6	358.8	363.0	367.2	371.5	375.8
22	380.1	384.4	388.8	393.2	397.6	402.0	406.4	410.9
23	415.4	420.0	424.5	429.1	433.7	438.3	443.0	447.6
24	452.3	457.1	461.8	466.6	471.4	476.2	481.1	485.9
25	490.8	495.7	500.7	505.7	510.7	515.7	520.7	525.8
26	530.9	536.0	541.1	546.3	551.5	556.7	562.0	567.2
27	572.5	577.8	583.2	588.5	593.9	599.3	604.8	610.2
28	615.7	621.2	626.7	632.2	637.9	643.5	649.1	654.8
29	660.5	666.2	671.9	677.7	683.4	689.2	695.1	700.9
30	706.8	712.7	718.6	724.6	730.6	736.6	742.6	748.6
31	754.8	760.9	767.0	773.1	779.3	785.5	791.7	798.0
32	804.3	801.6	816.9	823.2	829.6	836.0	842.4	848.8
33	855.3	861.8	868.3	874.9	881.4	888.0	894.6	901.3
34	907.9	914.7	921.3	928.1	934.8	941.6	948.4	955.3
35	962.1	969.0	975.9	982.8	989.8	996.8	1003.8	1010.8
36	1017.9	1025.0	1032.1	1039.2	1046.3	1053.5	1060.7	1068.0
37	1075.2	1082.5	1089.8	1097.1	1104.5	1111.8	1119.2	1126.7
38	1134.1	1141.6	1149.1	1156.6	1164.2	1171.7	1179.3	1186.9
39	1195.6	1202.3	1210.0	1217.7	1225.4	1233.2	1241.0	1248.8
40	1256.6	1264.5	1272.4	1280.3	1288.2	1296.2	1304.2	1312.2
41	1320.3	1328.3	1336.4	1344.5	1352.7	1360.8	1369.0	1377.2
42	1385.4	1393.7	1402.0	1410.3	1418.6	1427.0	1435.4	1443.8
43	1452.2	1460.7	1469.1	1477.6	1486.2	1494.7	1503.3	1511.9
44	1520.5	1529.2	1537.9	1546.6	1555.3	1564.0	1572.8	1581.6
45	1590.4	1599.3	1608.2	1617.0	1626.0	1634.9	1643.9	1652.9

THE LABOR QUESTION.

The vexed questions of wages, strikes and lock-outs, and demands for a reduced number of hours to constitute a day's
work, have at times to be met and decided. The world must
be taken as it is, not as any individual would have it. Trades'
Unions exist in the present as in the past, and are likely to increase in numbers and in power. Probably in no other country
have the rights of labor to a fair and equable share of the
profits of production been so fully recognized as in the United
States. A high protective tariff has for years received almost
unanimous public approval, chiefly because it was a shield
between the American workingman and the necessity which,
without protection, would have forced him to compete with the
overworked and underpaid labor of Europe. Manufacturers
generally have felt liberally disposed towards their *employées*
They desire that their men should live in comfortable houses,
wear good clothes, have plenty to eat, educate their children,
and accumulate something to live on when they are old.

Masters and men may be on the best of terms for years, when
suddenly some powerful Union gives the word of command
and a strike ensues. The argument of those who justify
strikes is that the masters are taking too large a share of the
gains of business and are giving the men too small a share.
The pretext is always the same ; either the masters protest that
profits have declined, and that a reduction of wages must follow, or the men allege that profits have augmented, and that an
increase of wages is reasonable. Evidently there is a great
deal to be said on both sides of these disputes.

The men do not and cannot know anything about the real
facts of business. It may be that their employers have quietly
continued paying for labor at a steady rate, through long
periods of continuous loss. Manufacturers who are located in
cities pay high rents and their workmen pay high rents, are

14

obliged to compete with others who manufacture in the country, where the workmen generally own the houses they live in, and where food is cheaper and dress plainer, and can therefore work cheaper. Home manufacturers have to compete with foreign manufacturers, who pay less wages and a lower rate of interest on capital. The employer has no power to impose upon the laborer, for if the latter is dissatisfied with his wages or his treatment he may go elsewhere or seek other employment. Neither the employer or *employé* can compel the other to pay or receive more or less than he is willing to give or take.

Nothing can control the price of labor but the law of supply and demand. If work is plenty laborers can increase their wages by demanding it, because the employer has no option. If work is scarce competition will bring down the price of labor, as it will of everything else. Each individual has the right to sell his labor for any price he can obtain for it, and any combination or organization designed to interfere with this right is against public policy and unlawful. A man may demand what he likes for his services, but his demand does not determine their value. If no one wants his services, a supply of anything, for which there is no demand, is valueless. It is of the utmost importance to an intelligent understanding of the relations of labor to capital that the workingman should appreciate the fact that society is not divided into two great antagonistic classes—capitalists and laborers. Every man who knows a trade, or is able to work, is as truly a capitalist as the man who owns a factory filled with costly machinery. His capital is his physical strength and his acquired skill in the performance of some useful labor. Labor is, and always will be merchandise. Those who have it to sell can only get for it so much as those who are asked to buy are willing to give. Self-interest, which is equally strong on both sides, operates to protect the seller against injustice and the buyer against extortion; while the public interest demands that the exchange of services should be free. If these simple elementary truths could be impressed upon the minds of workingmen, they would at once see the folly and futility of all efforts to artificially increase the value of their services. The natural laws of trade are as immutable in their operations to-day as they were centuries ago, and all human power cannot set them aside or suspend their operation. Intimidation, threats, or violence to persons or property, which have for their object a disturbance of the natural relations existing between labor and capital are crimes against society.

Trades' Unions, in many respects, are exceedingly beneficial. For members of a trade, working together in large numbers, who by their daily intercourse are made acquainted with each other's circumstances, and who are cognizant of much of the misery which is necessarily attendant on a precarious employment, would be inhuman indeed, if they did not unite for the purposes of mutual support in case of sickness, superannuation, for the burial of members and their wives, and also for assistance to members out of work. No one can look with disfavor on a Society organized for such a purpose. But when demagogues lead in what they delight to call " the war of labor upon capital," to elevate inferior workmen at the expense of superior skill ; to say that unionists shall not work with non-unionists ; that so many apprentices shall be allowed to a shop, and no more ; that so much work, and no more, and so many hours, and no more, shall be a day's service, and a system of terrorism practised to carry out these ideas—then a Society oversteps its useful purposes and its lawful rights.

The members of any particular trade, by earnestly uniting in the use of the various means of cultivation within their reach, may greatly increase the respectability and influence of that trade. Their funds, obtained by weekly contributions of members, will give security against the destitution which sickness may bring upon the most robust and industrious ; against the life-long dependence entailed by such calamities as disabling accidents, blindness, paralysis, or epilepsy, which incapacitate their victims from work ; and against the helplessness of old age, when failing powers render the continuous labor necessary to earn a livelihood impossible. A member may receive relief from the funds of his Society, and still maintain his self-respect. He has contributed to them in common with others. The man who is aware that when sickness or old age takes from him the power to labor, he will not be altogether deprived of a living, becomes a more contented as well as a more independent being.

Nearly all employers in this country have commenced life as workmen, and their places are to be filled again from the ranks. Can workmen who live by labor hope to secure more property by less labor ? In times past workmen have made demands that eight hours should constitute a day's work, and be paid the same wages as they had been accustomed to receive for ten hours' work. In New York city it is harder to labor ten hours a day than elsewhere, because a workingman to live respectably —away from a tenement house—must live at nearly an hour's distance from his shop. But it is neither wise nor just to ask the employers in one city to reduce the daily hours of labor to eight, while others are enjoying the advantages of ten hours. Now, suppose the change were made general, the result would be a reduction of one-fifth of the country's products in the necessaries and comforts of life ; and who would suffer most from this ? Not the wealthy men, whose money would secure what they required, but the workingmen, whose only source of income is their weekly wages. The working classes are consumers as well as producers, and share in the general benefits of the cheapening of the cost of the commodities they consume.

The condition of the working classes has greatly improved, and the improvement is still going on ; but it is an improvement which has taken place in spite of, and not because of, the lack of harmony between employers and employed. In ancient times the only energy employed in doing work for supplying man with the necessities or luxuries of life was that of muscular power, under a system of slavery. In modern times man has become in a great degree relieved from brute power, by substituting for his own muscular energy the power of nature, and this substitution is continually going on. The number of discoveries and improvements in the arts diminishes the amount of severe bodily labor. Education among mechanics multiplies these inventions ; and it should be a settled policy in every community to encourage in every possible way the intellectual cultivation of all who compose their body.

American mechanics are the most intelligent of any in the world, and with the most temperate social habits. They have a stimulus, in this country, of raising above their condition, or, at least, making it possible for their children to do so. The distributive industries—mercantile avocations—have long been, and probably always will be, overcrowded ; in the productive industries there always was, and probably always will be, plenty of room for mechanics who are thorough, honest, temperate and conscientious.

The loom, the sewing-machine, the steam-engine, the reaper, the printing-press, all increase the dignity and importance of mechanical labor. The multiplication of labor-saving machinery contributes to the desired attainment of universal abundance. The progress toward abundance must necessarily be slow, however active production may be. The proportion of people, even in this favored land, who have reached the condition in which they can say that all their reasonable wants and desires are satisfied, and that they enjoy abundance, is certainly very small. A vast number have never known what it is to have had enough of food and clothing. With so great a void yet to be filled there can be no such thing as over-production. It is to a still higher development, and a yet more general employment of labor-saving machinery, that must bring more general prosperity in the future. Experience has shown that while machinery increases production, it also opens new fields for useful labor. Then, too, the cheapening of the cost of manufactured products, proportionately increases their consumption, by bringing them within the reach of a greater number of persons. Workingmen need never fear from the introduction of labor-saving machinery. To point to it in fear of an over-production, and consequent enforced idleness of skilled labor, indicates a short-sightedness as great as that which impelled the French silk weavers to destroy the loom of Jacquard, which, instead of taking away the work of a few hundred half-starved, consumptive workmen, has given employment to an army of well-fed, well-clothed, and comfortably housed operatives, and has added uncounted millions to the world's wealth.

The efforts which will be attended with the most encouraging results between master and men, are those which seek to reduce the cost of living. The workingman is not usually so situated that he can purchase anything to advantage. Where manufacturers are so situated that they can do so, they should provide small, neat and convenient houses for their *employés*. and rent them for just enough to cover interest, taxes, and repairs. A few items of living necessaries, such as coal, flour, etc., should be provided for cash sales at prime cost. The object of this is to enable the workingman to get the greatest possible value for his money, in order that he may be able to live well for the smallest wages. The expense and trouble is a mere nothing, and the gratitude of the men out of all proportion to the work rendered. By increasing the purchasing power of their wages, so far as practicable, good feeling arises between the employed and the employer, and the much desired alliance, offensive and defensive, between labor and capital, becomes not only possible, but extremely probable.

TO YOUNG MEN.

CAPABLE men to manage Architectural Iron Works are scarce. Looking over the field—the enormous business that must certainly be done in every part of the Union—the coming demand for the right kind of men will be greater than the supply. Of men of mediocrity there will always be an abundance. So many requirements go to fill the bill, that first-class men will always be in demand. The foundation must be a natural talent and liking for mechanical work. No one can succeed if incapacitated by disposition or education. The toil must be congenial. A boy who has given evidence of ingenuity and dexterity with the use of tools, can make choice of this pursuit in life, with the certainty of eventually reaping pecuniary independence and a happy and honorable career. Success in this business depends upon fitness for undertaking it, coupled with conscientious labor. No more honorable or profitable profession or business can be selected. In a country like ours, a claim of superior respectability on behalf of any calling is preposterous; the circumstance of being an American citizen is sufficient to adorn with all proper dignity any trade or profession which a young man may adopt. In point of real and essential respectability, all trades and professions are equal; and the social position which a man enjoys, and the degree of respect which he is able to command, depend not upon his trade, but upon his individual character. Thousands of young men have entered the learned professions when they were already crowded, and are consequently wasting their lives in vain hopes; and other thousands have devoted themselves to the pursuits of commerce without capital, prudence, or intelligence sufficient to avoid the dangers of commercial enterprise, to become either bankrupts or involved in a series of embarrassments which may last through their whole lives. An error in the choice of one's profession is one which is followed by painful consequences, as many have found to their cost.

Having made a choice of this business, and possessed of a good common school education, there must follow some years of practical learning.   First should come an apprenticeship of not less than two years, with an architect of large practice, so as to become familiar with the plans and constructions of buildings generally, the making of detail drawings, and the way and manner of doing things generally in such an office.   Then in the shop: one year in the pattern shop; two years in the foundry, learning to become a moulder; and two years after that as a finisher, in fitting up cast-iron work, and doing wrought-iron and blacksmith work.   These seven years of continuous daily toil will be happy years.   At their expiration, the man—it is to be hoped a gentleman, withal—will be fitted to take off quantities from plans, to make estimates and secure contracts, and superintend with intelligence and authority the workmen under his care.   With the age of manhood, the heavy duties and fearful responsibilities of active life will come to him when his judgment is matured, his understanding ripened, and his nerves hardened for the rough encounter of conflicting interests and unforeseen emergencies.

On his energy, perseverance and skill, will depend how large a sphere he will fill.   It all depends on himself.   If inspired by an honest ambition to excel, and willing to study the literature which modern book-making has placed so easily within his reach, his chances of success in life are far more numerous and certain than those of any other class of young men in the community.   The business openings will be sufficiently numerous to satisfy the largest ambition.

It is of great importance that his leisure time be given to the cultivation of his mind.   If the physician, the lawyer and the divine avail themselves of the assistance of science and literature in their several professions, the mechanic has still stronger inducements for doing the same thing; for, to none of these professions are the results of science so directly applicable, and for none of them are the recreations of literature so appropriate or gratifying. By making himself master of those principles which are most intimately connected with Architectural Iron Work, he, while satisfying a liberal curiosity, may possibly be approaching some brilliant discovery which will speedily conduct him to fortune and fame. Each of the mechanical trades affords ample room for the exercise of ingenuity in the improvement of its processes, and the consequent improvement of its products. Abundant trade periodicals exist, journals devoted to architecture and building, to engineering, to the iron interest ; from a thousand sources ideas are to be got.

It is not desirable for a man to devote every moment of his time to the business by which he lives. Such intense application is injurious both to the body and the mind. It destroys health, racks the brain, and ruins the temper. The repose of the domestic circle, the quiet hour for reading, or relaxation of some other kind, seem absolutely necessary for the preservation of that greatest of earthly blessings—a sound mind in a healthy body. A mechanical business, a life of activity and labor, is far from being unfavorable to the highest operations of the intellect; and that relaxation from active labors is most appropriately found in mental recreations. Whether, therefore, he addresses himself to increasing the quantity or improving the quality of his manufactures, the paths before him are wide enough for his greatest powers and his most unwearied activity. Let him remember that knowledge is power, and neglect no opportunity of improving his mind. Seasons of depression may affect more or less such a man, but these cannot rob him of his capital gained at the bench, the drawing table and the evening fireside. There is no class of men so absolutely independent of chance and mischance, as mechanics. If more of our intelligent young men, with good educations and good social positions, would learn the various mechanical trades, fewer of them would have occasion in after years to bemoan the wasted opportunities of youth, and the fruitless struggles of an unsuccessful life.

If a young man has received a college education it is well. But let it be understood, that those who utterly lack in high scholarship, have the same open road to an honorable, useful and independent career. Indeed, it is better that the practical art precede the science. One great thing needed, that cannot be learned in school, is how to deal with men; how to make them work in accordance with your ideas. You may make a perfect plan, and have a complete drawing, but if you cannot impress it upon your foreman of pattern makers, it will not prove a success. And so in every department. The difference between a beautiful line and one which has no beauty whatever, is very frequently a mere nothing—so undefinedly small that one can scarcely say in what the difference consists. If you have the skill to add the finishing touches to a set of patterns, or take a file and clean up a part that needs but a touch to make it perfection, you will not only make an admirable workman, but will do much toward a high standard of work in the shop. Some of the finest pieces of wrought iron work extant were designed by men who blew the bellows and swung the hammer; and the same may be said of some of the best examples of cast iron art work. The value of many manufactures is chiefly due to their beauty. There is hardly any limit to the market value of beauty—that element in manufactures which responds to the finer sensibilities of man. Not only the methods of working, but the nature and capabilities of materials must be understood. A design that would be admirable in silver would in all probability be hideous in cast-iron. The quality of an article may be said to consist of three principal elements: 1st, adaptation to the purpose for which the object was made; 2d, durability; 3d, beauty. For instance, a column should certainly possess the first two elements; and, all other things being equal, every builder will sooner pay his money for a handsome column than for a homely one. He may not be willing to pay an additional dollar for a column, simply because of its beauty; but since it is as cheap for an iron man, who has good taste, to give his column a certain degree of comeliness, as to make it atrociously ugly, he finds his profit in a readier sale.

No matter what amount of mental culture a man brings to this business, he cannot bring too much. If an apprentice be lacking in certain kinds of book knowledge, he can and must acquire it. The practical man separates useless stuff from that which is valuable, and he can more easily acquire knowledge on special subjects, than can the school-man acquire practical knowledge of tools and machinery, the management of labor, and the general principles of economy in construction, maintenance and working. Each is possessed of certain knowledge that the other must learn; and years of study, years of labor, will make them equal. The one having the most energy and perseverance will prove the better man.

Executive ability, business tact, and good management in finances—these come after an experience of the annoyances, anxieties, discomforts and sufferings inevitable to business life. As a matter of policy, as well as of duty, an upright, moral life —ever truthful and strictly honest—is the best. " For what is a man profited, if he shall gain the whole world, and lose his own soul ? "

This short chapter is for the kindly encouragement of young men who desire, or are about to follow this branch of business, or are actually engaged therein. I would especially caution young men not to care for the ill-natured remarks that may be said of them. Jealousy, envy and malice will pursue always, and to the sensitive cut to the heart. It takes many years to become calloused. In proportion to your ability, trade slanderers will pursue, and a courageous heart is necessary in the fight. Bide your time patiently; the turn of Fortune's wheel brings many changes. Earn a reputation for reliability as to word and promise, and for secrecy in confidential matters. Mind your own business, and treat all men in accordance with the golden rule. Confide in your own strength without boasting of it ; respect that of others without fearing it. Have enthusiasm in your calling, faith in the future, intelligence in your work, endurance and an unconquerable will.

LOSE NO TIME.

JOHN WILEY & SONS'

LIST OF PUBLICATIONS,

15 ASTOR PLACE.

*Books marked with an * are sold at net prices to the Trade.*

AGRICULTURE.

DOWNING.   **FRUITS AND FRUIT-TREES OF AMERICA;** or the Culture, Propagation, and Management in the Garden and Orchard, of Fruit-trees generally, with descriptions of al. the finest varieties of Fruit, Native and Foreign, cultivated in this country. By A. J. Downing. Second revision and correction, with large additions. By Chas. Downing. 1 vol. 8vo. over 1100 pages, with several hundred outline engravings. Price, with Supplement for 1872.........................$5 00

"As a work of reference it has no equal in this country, and deserves a place in the Library of every Pomologist in America."—*Marshall P. Wilder.*

"   **ENCYCLOPEDIA OF FRUITS;** or, Fruits and Fruit-Trees of America. Part 1.—APPLES. With an Appendix containing many new varieties, and brought down to 1872. By Chas. Downing. With numerous outline engravings. 8vo, full cloth...$2 50

"   **ENCYCLOPEDIA OF FRUITS;** or, Fruits and Fruit-Trees of America. Part 2.—CHERRIES, GRAPES, PEACHES, PEARS, &c. With an Appendix containing many new varieties, and brought down to 1872. By Chas. Downing. With numerous outline engravings. 8vo, full cloth.........$2 50

*   **FRUITS AND FRUIT-TREES OF AMERICA.** By A. J. Downing. First revised edition. By Chas. Downing 12mo, cloth...

*   **SELECTED FRUITS.** From Downing's Fruits and Fruit-Trees of America. With some new varieties, including their Culture. Propagation, and Management in the Garden and Orchard, with a Guide to the selection of Fruits, with reference to the Time of Ripening. By Chas. Downing. Illustrated with upwards of four hundred outlines of Apples, Cherries, Grapes, Plums, Pears, &c. 1 vol., 12mo....$2 50

"   **LOUDON'S GARDENING FOR LADIES, AND COMPANION TO THE FLOWER-GARDEN.** Second American from third London edition. Edited by A. J. Downing. 1 vol., 12mo $2 00

DOWNING & LINDLEY.   **THE THEORY OF HORTICULTURE.** By J. Lindley. With additions by A. J. Downing. 12mo, cloth.......$2 00

DOWNING.   **COTTAGE RESIDENCES.** A Series of Designs for Rural Cottages and Cottage Villas, with Garden Grounds. By A. J. Downing. Containing a revised List of Trees, Shrubs, and Plants, and the most recent and best selected Fruit, with some account of the newer style of Gardens. By Henry Winthrop Sargent and Charles Downing. With many new designs in Rural Architecture. By George E Harney, Architect. 1 vol. 4to.........................$6 00

DOWNING & WIGHTWCK. **HINTS TO PERSONS ABOUT BUILDING IN THE COUNTRY.** By A. J. Downing. And **HINTS TO YOUNG ARCHITECTS,** calculated to facilitate their practical operations. By George Wightwick, Architect. Wood engravings. 8vo, cloth.................$2 00

KEMP. **LANDSCAPE GARDENING;** or, How to Lay Out a Garden. Intended as a general guide in choosing, forming, or improving an estate (from a quarter of an acre to a hundred acres in extent), with reference to both design and execution. With numerous fine wood engravings. By Edward Kemp. 1 vol. 12mo, cloth........................$2 50

LIEBIG **CHEMISTRY IN ITS APPLICATION TO AGRICULTURE,** &c. By Justus Von Liebig. 12mo, cloth....$1 00

" **LETTERS ON MODERN AGRICULTURE.** By Baron Von Liebig. Edited by John Blyth, M.D. With addenda by a practical Agriculturist, embracing valuable suggestions, adapted to the wants of American Farmers. 1 vol. 12mo, cloth...$1 00

" **PRINCIPLES OF AGRICULTURAL CHEMISTRY,** with special reference to the late researches made in England. By Justus Von Liebig. 1 vol. 12mo.................75 cents.

PARSONS. **HISTORY AND CULTURE OF THE ROSE.** By S. B. Parsons. 1 vol. 12mo.............................$1 25

ARCHITECTURE.

DOWNING. **COTTAGE RESIDENCES;** or, a Series of Designs for Rural Cottages and Cottage Villas and their Gardens and Grounds, adapted to North America. By A. J. Downing. Containing a revised List of Trees, Shrubs, Plants, and the most recent and best selected Fruits. With some account of the newer style of Gardens, by Henry Wentworth Sargent and Charles Downing. With many new designs in Rural Architecture by George E. Harney, Architect........................$6 00

DOWNING & WIGHTWICK. **HINTS TO PERSONS ABOUT BUILDING IN THE COUNTRY.** By A. J. Downing. And **HINTS TO YOUNG ARCHITECTS,** calculated to facilitate their practical operations. By George Wightwick, Architect. With many wood-cuts. 8vo, cloth..................$2 00

HATFIELD. **THE AMERICAN HOUSE CARPENTER.** A Treatise upon Architecture, Cornices, and Mouldings, Framing, Doors, Windows, and Stairs; together with the most important principles of Practical Geometry. New, thoroughly revised, and improved edition, with about 150 additional pages, and numerous additional plates. By R. G. Hatfield. 1 vol. 8vo..$3 50

NOTICES OF THE WORK.
" The clearest and most thoroughly practical work on the subject."
" This work is a most excellent one, very comprehensive, and lucidly arranged."
" This work commends itself by its practical excellence."
" It is a valuable addition to the library of the architect, and almost indispensable to every scientific master-mechanic."—*R. R. Journal.*

HOLLY **CARPENTERS' AND JOINERS' HAND-BOOK,** containing a Treatise on Framing, Roofs, etc., and useful Rules and Tables. By H. W. Holly. 1 vol. 18mo, cloth........$0 75

" **THE ART OF SAW-FILING SCIENTIFICALLY TREATED AND EXPLAINED.** With Directions for putting in order all kinds of Saws. By H. W. Holly. 18mo, cloth.......................................$0 75

RUSKIN **SEVEN LAMPS OF ARCHITECTURE.** 1 vol. 12mo, cloth, plates.... ....$1 75

RUSKIN **LECTURES ON ARCHITECTURE AND PAINTING** 1 vol. 12mo, cloth, plates...........................$1 50

" **LECTURE BEFORE SOCIETY OF ARCHITECTS.** 0 15

WOOD. **A TREATISE ON THE RESISTANCE OF MA-TERIALS,** and an Appendix on the Preservation of Timber By De Volson Wood, Prof. of Engineering. University of Michigan. 2d edition, thoroughly revised. 8vo, cloth. $3 00

This work is used as a Text-Book in Iowa University, Iowa Agricultural College, Illinois Industrial University, Sheffield Scientific School, New Haven, Cooper Institute, New York, Polytechnic College, Brooklyn, University of Michigan. and other institutions.

" **A TREATISE ON BRIDGES.** Designed as a Text-book and for Practical Use. By De Volson Wood. 1 vol. 8vo, nu merous illustrations, ...$3 00

ASSAYING—ASTRONOMY.

BODEMANN. **A TREATISE ON THE ASSAYING OF LEAD, SILVER, COPPER, GOLD, AND MERCURY.** By Bodemann and Kerl. Translated by W. A. Goodyear. 1 vol. 12mo, cloth $2 50

MITCHELL. **A MANUAL OF PRACTICAL ASSAYING.** By John Mitchell. Fourth edition, edited by William Crookes. 1 vol. thick 8vo, cloth...........................$10 00

NORTON. **A TREATISE ON ASTRONOMY, SPHERICAL AND PHYSICAL,** with Astronomical Problems and Solar, Lunar, and other Astronomical Tables for the use of Colleges and Scientific Schools. By William A. Norton. Fourth edition, revised, remodelled, and enlarged. Numerous plates. 8vo, cloth $3 50

BIBLES, &c.

BAGSTER. **THE COMMENTARY WHOLLY BIBLICAL.** Contents: —The Commentary: an Exposition of the Old and New Testaments in the very words of Scripture. 2264 pp. II. An outline of the Geography and History of the Nations mentioned in Scripture. III. Tables of Measures, Weights, and Coins. IV. An Itinerary of the Children of Israel from Egypt to the Promised Land. V. A Chronological comparative Table of the Kings and Prophets of Israel and Judah. VI. A Chart of the World's History from Adam to the Third Century, A. D. VII. A complete Series of Illustrative Maps. IX. A Chronological Arrangement of the Old and New Testaments. X. An Index to Doctrines and Subjects, with numerous Selected Passages, quoted in full. XI. An Index to the Names of Persons mentioned in Scripture. XII. An Index to the Names of Places found in Scripture. XIII. The Names, Titles, and Characters of Jesus Christ our Lord, as revealed in the Scriptures, methodically arranged.

2 volumes 4to, cloth.............................$19 50
2 volumes 4to, half morocco, gilt edges............. 26 00
2 volumes 4to, morocco, gilt edges.......... .... 35 00
3 volumes 4to, cloth............................. 20 00
3 volumes 4to, half morocco, gilt edges............. 33 00
3 volumes 4to, morocco, gilt edges................. 40 00

BLANK-PAGED BIBLE. **THE HOLY SCRIPTURES OF THE OLD AND NEW TESTAMENTS;** with copious references to parallel and illustrative passages, and the alternate pages ruled for MS. notes.

This edition of the Scriptures contains the Authorized Version, illustrated by the references of "Bagster's Polyglot Bible," and enriched with accurate maps, useful tables, and an Index of Subjects.

1 vol. 8vo, morocco extra.........................$9 50
1 vol. 8vo, full morocco.........................11 00

*THE TREASURY BIBLE. Containing the authorized English version of the Holy Scriptures, interleaved with a Treasury of more than 500,000 Parallel Passages from Canne, Brown, Blayney, Scott, and others. With numerous illustrative notes.

1 vol., half bound.................................$7 50
1 vol., morocco.................................10 00

COMMON PRAYER, 48mo Size.
(Done in London expressly for us.)

COMMON PRAYER. No. 1. Gilt and red edges, imitation morocco..........$0 62½
No. 2. Gilt and red edges, rims....................... 87½
No. 3. Gilt and red edges, best morocco and calf....... 1 25
No. 4. Gilt and red edges, best morocco and calf. rims.. 1 50

BOOK-KEEPING.

JONES. **BOOKKEEPING AND ACCOUNTANTSHIP.** Elementary and Practical. In two parts, with a Key for Teachers. By Thomas Jones, Accountant and Teacher. 1 volume 8vo, cloth...$2 50

" **BOOKKEEPING AND ACCOUNTANTSHIP.** School Edition. By Thomas Jones. 1 vol. 8vo, half roan......$1 50

" **BOOKKEEPING AND ACCOUNTANTSHIP.** Set of Blanks. In 6 parts. By Thomas Jones..............$1 50

" **BOOKKEEPING AND ACCOUNTANTSHIP.** Double Entry; Results obtained from Single Entry; Equation of Payments, etc. By Thomas Jones. 1 vol. thin 8vo...$0 75

CHEMISTRY.

CRAFTS. **A SHORT COURSE IN QUALITATIVE ANALYSIS;** with the new notation. By Prof. J. M. Crafts. Second edition. 1 vol. 12mo, cloth........................$1 50

JOHNSON'S FRESENIUS. **A MANUAL OF QUALITATIVE CHEMICAL ANALYSIS.** By C. R. Fresenius. Translated into the New System, and newly edited by Samuel W. Johnson, M.A., Prof. of Theoretical and Agricultural Chemistry, in the Sheffield Scientific School of Yale College, New Haven. 1 vol. 8vo, cloth. 1875.................................$4 50

DITTO. Edition of 1874. 8vo, cloth.............$3.00

" **A SYSTEM OF INSTRUCTION IN QUANTITATIVE CHEMICAL ANALYSIS.** By C. R. Fresenius. From latest editions, edited, with additions, by Prof. S. W. Johnson. With Chemical Notation and Nomenclature, old and new..$6 00

KIRKWOOD **COLLECTION OF REPORTS (CONDENSED) AND OPINIONS OF CHEMISTS IN REGARD TO THE USE OF LEAD PIPE FOR SERVICE PIPE,** in the Distribution of Water for the Supply of Cities. By Jas. P. Kirkwood. 8vo, cloth...........................$1 50

MILLER. **ELEMENTS OF CHEMISTRY, THEORETICAL AND PRACTICAL.** By Wm. Allen Miller. 3 vols. 8vo..$18 00

" Part I.—CHEMICAL PHYSICS. 1 vol. 8vo..........$4 00

" Part II.—INORGANIC CHEMISTRY. 1 vol. 8vo..... 6 00

" Part III.—ORGANIC CHEMISTRY. 1 vol. 8vo.......10 00

"Dr. Miller's Chemistry is a work of which the author has every reason to feel proud. It is now by far the largest and most accurately written Treatise on Chemistry in the English language," etc.—*Dublin Med. Journal.*

" **MAGNETISM AND ELECTRICITY.** By Wm. Allen Miller. 1 vol. 8vo...$2 50

*MUSPRATT.	**CHEMISTRY — THEORETICAL, PRACTICAL, AND ANALYTICAL**—as applied and relating to the Arts and Manufactures. By Dr. Sheridan Muspratt. 2 vols. 8vo, cloth, $19.00; half russia.........................$25 00
PERKINS.	**AN ELEMENTARY MANUAL OF QUALITATIVE CHEMICAL ANALYSIS.** By Maurice Perkins. 12mo, cloth...$1 00
THORPE.	**QUANTITATIVE CHEMICAL ANALYSIS.** By T. E. Thorpe, Prof. of Chemistry, Glasgow. 1 vol. 18mo, plates. Cloth.................... ... $1 75

Prof. S. W. Johnson says of this work:— "I know of no other small book of anything like its value."

"This very excellent and orginal work has long been waited for by scientific men."—*Scientific American.*

DRAWING AND PAINTING.

BOUVIER AND OTHERS.	**HANDBOOK ON OIL PAINTING.** Handbook of Young Artists and Amateurs in Oil Painting; being chiefly a condensed compilation from the celebrated Manual of Bouvier, with additional matter selected from the labors of Merriwell, De Montalbert, and other distinguished Continental writers on the art. In 7 parts. Adapted for a Text-Book in Academies of both sexes, as well as for self-instruction Appended, a new Explanatory and Critical Vocabulary. By an American Artist. 12mo, cloth...................$2 00
COE.	**PROGRESSIVE DRAWING BOOK.** By Benj. H. Coe. One vol., cloth...............................$3 50
"	**DRAWING FOR LITTLE FOLKS**; or, First Lessons for the Nursery. 30 drawings. Neat cover...........$0 20
"	**FIRST STUDIES IN DRAWING.** Containing Elementary Exercises, Drawings from Objects, Animals, and Rustic Figures. Complete in *three numbers* of 18 studies each, in neat covers. Each........................$0.20
"	**COTTAGES.** An Introduction to Landscape Drawing. *Containing 72 Studies.* Complete in four numbers of 18 studies each, in neat covers. Each... ....$0.20
"	**EASY LESSONS IN LANDSCAPE.** Complete in four numbers of 10 Studies each. In neat 8vo cover. Each, $0 20
"	**HEADS, ANIMALS, AND FIGURES.** Adapted to Pencil Drawing. Complete in three numbers of 10 Studies each. In neat 8vo covers. Each... ....$0 20
"	**COPY BOOK, WITH INSTRUCTIONS**............$0 37½
MAHAN.	**INDUSTRIAL DRAWING.** Comprising the Description and Uses of Drawing Instruments, the Construction of Plane Figures, the Projections and Sections of Geometrical Solids, Architectural Elements, Mechanism, and Topographical Drawing. With remarks on the method of Teaching the subject. For the use of Academies and Common Schools. By Prof. D. H. Mahan. 1 vol. 8vo. Twenty steel plates. Full cloth......................................$3 00
RUSKIN.	**THE ELEMENTS OF DRAWING.** In Three Letters to Beginners. By John Ruskin. 1 vol. 12mo..........$1 00
"	**THE ELEMENTS OF PERSPECTIVE.** Arranged for the use of Schools. By John Ruskin...................$1 00
SMITH.	**A MANUAL OF TOPOGRAPHICAL DRAWING.** By Prof. R. S. Smith. New edition with additions. 1 vol. 8vo, cloth, plates............................$2.00
"	**MANUAL OF LINEAR PERSPECTIVE.** Form, Shade, Shadow, and Reflection. By Prof. R. S. Smith. 1 vol. 8vo, plates, cloth.............................$2 00
WARREN.	1. **ELEMENTARY FREE-HAND GEOMETRICAL DRAWING.** A series of progressive exercises on regular lines and forms, including systematic instruction in lettering; a training of the eye and hand for all who are learning to draw. 12mo, cloth, many cuts............................. 75 cts. DITTO, including *Drafting Instruments,* etc. 12mo, cl..$1 75

6

ELEMENTARY WORKS.—Continued.

WARREN

2. PLANE PROBLEMS IN ELEMENTARY GEOMETRY. With numerous wood-cuts. 12mo, cloth................. $1 20

3. DRAFTING INSTRUMENTS AND OPERATIONS. Containing full information about all the instruments and materials used by the draftsmen, with full directions for their use. With plates and wood-cuts. One vol. 12mo, cloth, $1 25

4. ELEMENTARY PROJECTION DRAWING. Revised and enlarged edition. In five divisions. This and the last volume are favorite text-books, especially valuable to all Mechanical Artisans, and are particularly recommended for the use of all higher public and private schools. New revised and enlarged edition, with numerous wood-cuts and plates. (1872.) 12mo, cloth $1 50

5. ELEMENTARY LINEAR PERSPECTIVE OF FORMS AND SHADOWS. Part I.—Primitive Methods, with an Introduction. Part II.—Derivative Methods, with Notes on Aerial Perspective, and many Practical Examples. Numerous wood-cuts. 1 vol. 12mo, cloth..........................$1 00

II. HIGHER WORKS.

These are designed principally for Schools of Engineering and Architecture, and for the members generally of those professions; and the first three are also designed for use in those colleges which provide courses of study adapted to the preliminary general training of candidates for the scientific professions, as well as for those technical schools which undertake that training themselves.

1. DESCRIPTIVE GEOMETRY, OR GENERAL PROBLEMS OF ORTHOGRAPHIC PROJECTIONS. The foundation course for the subsequent theoretical and practical works. 1 vol. 8vo, 24 folding plates and woodcuts...........$3 50

2. GENERAL PROBLEMS OF SHADES AND SHADOWS. A wider range of problems than can elsewhere be found in English, and the principles of shading. 1 vol. 8vo, with numerous plates. Cloth......................... $3 00

3. HIGHER LINEAR PERSPECTIVE. Distinguished by its concise summary of various methods of perspective construction; a full set of standard problems, and a careful discussion of special higher ones. With numerous large plates. 8vo, cloth...$3 50

4. ELEMENTS OF MACHINE CONSTRUCTION AND DRAWING; or, Machine Drawings. With some elements of descriptive and rational cinematics. A Text-Book for Schools of Civil and Mechanical Engineering, and for the use of Mechanical Establishments, Artisans, and Inventors. Containing the principles of gearings, screw propellers, valve motions, and governors, and many standard and novel examples, mostly from present American practice. By S. Edward Warren. 2 vols. 8vo. 1 vol. text and cuts, and 1 vol. large plates.... . $7 50

STONE CUTTING. A Treatise on the Graphics and Practice of Stone Cutting, for Engineers, Architects, Masons, and Students. 1 vol. 8vo, plates $2 50

A FEW FROM MANY TESTIMONIALS.

"It seems to me that your Works only need a thorough examination to be introduced and permanently used in all the Scientific and Engineering Schools."
—Prof. J. G. FOX, *Collegiate and Engineering Institute, New York City.*

"I have used several of your Elementary Works, and believe them to be better adapted to the purposes of instruction than any others with which I am acquainted."—H. F. WALLING, *Prof. of Civil and Topographical Engineering, Lafayette College, Easton, Pa.*

"The author has happily divided the subjects into two great portions: the former embracing those processes and problems proper to be taught to all students in Institutions of Elementary Instruction; the latter, those suited to advanced students preparing for technical purposes. The Elementary Books ought to be used in all High Schools and Academies; the Higher ones in Schools of Technology."—WM. W. FOLWELL, *President of University of Minnesota.*

DYEING, &c.

CALVERT. **DYEING AND CALICO PRINTING.** By C. Calvert. Edited by Dr. Stenhouse and C. E. Groves. Illustrated with wood engravings and specimens of printed and dyed fabrics. (Ready in October.) 1 vol. 8vo....................$8 00

MACFARLANE. **A PRACTICAL TREATISE ON DYEING AND CALICO PRINTING.** Including the latest Inventions and Improvements. With an Appendix, comprising definitions of chemical terms, with tables of Weights, Measures, &c. By an experienced Dyer. With a supplement, containing the most recent discoveries in color chemistry. By Robert Macfarlane. 1 vol. 8vo...$5 00

REIMANN **A TREATISE ON THE MANUFACTURE OF ANILINE AND ANILINE COLORS.** By M. Reimann. To which is added the Report on the Coloring Matters derived from Coal Tar, as shown at the French Exhibition, 1867. By Dr. Hofmann. Edited by Wm. Crookes. 1 vol. 8vo, cloth, $2 50

"Dr. Reimann's portion of the Treatise, written in concise language, is profoundly practical, giving the minutest details of the processes for obtaining all the more important colors, with woodcuts of apparatus. Taken in conjunction with Hofmann's Report, we have now a complete history of Coal Tar Dyes, both theoretical and practical."—*Chemist and Druggist.*

ENGINEERING.

AUSTIN. **A PRACTICAL TREATISE ON THE PREPARATION, COMBINATION, AND APPLICATION OF CALCAREOUS AND HYDRAULIC LIMES AND CEMENTS.** To which is added many useful recipes for various scientific, mercantile, and domestic purposes. By James G. Austin. 1 vol. 12mo.......................................$2 00

COLBURN **LOCOMOTIVE ENGINEERING AND THE MECHANISM OF RAILWAYS.** A Treatise on the Principles and Construction of the Locomotive Engine, Railway Carriages, and Railway Plant, with examples. Illustrated by Sixty-four large engravings and two hundred and forty woodcuts. By Zerah Colburn. Complete, 20 parts, $15.00; or 2 vols. cloth.....................................$16 00
Or, half morocco, gilt top........................$20 00

DU BOIS. **1. ELEMENTS OF GRAPHICAL STATICS,** and their Application to Framed Structures, etc. Cranes; Bridge, Roof, and Suspension Trusses; Braced and Stone Arches; Pivot and Draw Spans; Continuous Girders, etc. By A. J. Du Bois, C.E., Ph.D. 2 vols. 8vo, 1 vol. text and 1 vol. plates.............................. ....$5 00

HERSCHEL **2. A HANDBOOK FOR BRIDGE ENGINEERS.** By C. Herschel. In 3 vols. Each vol. complete in itself. Vol. I. Straight and Beam Bridges. Vol. II. Suspension and Arched Bridges. Vol. III. Stone Bridges; Bridge Piers and their Foundations.

MAHAN. **AN ELEMENTARY COURSE OF CIVIL ENGINEERING,** for the use of the Cadets of the U. S. Military Academy. By D. H. Mahan. 1 vol. 8vo, with numerous illustrations, and an Appendix and general Index. Edited by Prof. De Volson Wood. Full cloth.........................$5 00

" **DESCRIPTIVE GEOMETRY, as** applied to the Drawing of Fortifications and Stone-Cutting. For the use of the Cadets of the U. S. Military Academy. By Prof. D. H. Mahan. 1 vol. 8vo. Plates...............................$1 50

" **A TREATISE ON FIELD FORTIFICATIONS.** Containing instructions on the Methods of Laying Out, Constructing, Defending, and Attacking Entrenchments. With the General Outlines, also, of the Arrangement, the Attack, and Defence of Permanent Fortifications. By Prof. D. H. Mahan. New edition, revised and enlarged. 1 vol. 8vo, full cloth, with plates..$3 50

" **ELEMENTS OF PERMANENT FORTIFICATIONS.** By Prof. D. H. Mahan. 1 vol. 8vo, with numerous large plates. Revised and edited by Col. J. B. Wh $6 50

MAHAN.   **ADVANCED GUARD, OUT-POST**, and Detachment Service of Troops, with the Essential Principles of Strategy and Grand Tactics. For the use of Officers of the Militia and Volunteers. By Prof. D. H. Mahan. New edition, with large additions and 12 plates. 1 vol. 18mo, cloth......$1 50

MAHAN & MOSELY.   **MECHANICAL PRINCIPLES OF ENGINEERING AND ARCHITECTURE.** By Henry Mosely, M.A., F.R.S. From last London edition, with considerable additions, by Prof. D. H. Mahan, LL.D., of the U. S. Military Academy. 1 vol. 8vo, 700 pages. With numerous cuts. Cloth...$5 00

MAHAN & BRESSE.   **HYDRAULIC MOTORS.** Translated from the French Cours de Mecanique, appliquée par M. Bresse. By Lieut. F. A. Mahan, and revised by Prof. D. H. Mahan. 1 vol. 8vo, plates. 1876...$2 50

WOOD.   **A TREATISE ON THE RESISTANCE OF MATERIALS,** and an Appendix on the Preservation of Timber. By De Volson Wood, Professor of Engineering, University of Michigan. 1 vol. 8vo, cloth..................... ....$3 00

     A TREATISE ON BRIDGES. Designed as a Text-book and for Practical Use. By De Volson Wood. 1 vol. 8vo, numerous illustrations, cloth $3 00

GREEK.

BAGSTER.   **GREEK TESTAMENT, ETC.** The Critical Greek and English New Testament in Parallel Columns, consisting of the Greek Text of Scholz, readings of Griesbach, etc., etc. 1 vol. 18mo, half morocco....................$2 50

"   —— do.   Full morocco, gilt edges................ 4 50

"   —— With Lexicon, by T. S. Green. Half-bound...... 4 00

"   —— do. Full morocco, gilt edges......... 6 00

"   **GREEK TESTAMENT.** Lexicon and Concordance. Half-bound..$5 00

"   —— Morocco limp, $6.50; morocco flaps, $7.00; morocco flaps, calf lined............................. 7 50

"   **THE ANALYTICAL GREEK LEXICON TO THE NEW TESTAMENT.** In which, by an alphabetical arrangement, is found every word in the Greek text *in every form in which it appears*—that is to say, every occurrent person, number, tense or mood of verbs, every case and number of nouns, pronouns, &c., is placed in its alphabetical order, fully explained by a careful grammatical analysis and referred to its root. 1 vol. small 4to, half-bound...................... ...$6 50

"   **GREEK TESTAMENT.** By Griesbach and Greenfield. 32mo. Half-bound...... ....$1 75

"   DITTO. With Lexicon. 32mo, half-bound.............$2 25

GREENFIELD   **GREEK LEXICON** (Polymicrian.) 32mo, half.bound..$1 00

"   **GREEK-ENGLISH LEXICON TO TESTAMENT.** By T. S. Green. Half morocco $1 50

HEBREW.

GREEN.   **A GRAMMAR OF THE HEBREW LANGUAGE.** With copious Appendixes. By W. H. Green, D.D., Professor in Princeton Theological Seminary. 1 vol. 8vo, cloth....$3 50

"   **AN ELEMENTARY HEBREW GRAMMAR.** With Tables, Reading Exercises, and Vocabulary. By Prof. W. H. Green, D.D. 1 vol. 12mo, cloth....................$1 25

"   **HEBREW CHRESTOMATHY;** or, Lessons in Reading and Writing Hebrew. By Prof. W. H. Green, D.D. 1 vol. 8vo, cloth...$2 00

*LETTERIS   **A NEW AND BEAUTIFUL EDITION OF THE HEBREW BIBLE.** Revised and carefully examined by Myer Levi Letteris. 1 vol. 8vo, with key, marble edges.....$2 50

"This edition has a large and much more legible type than the known one volume editions, and the print is excellent, while the name of LETTERIS is a sufficient guarantee for correctness."—*Rev. Dr. J. M. WISE, Editor of the ISRAELITE.*

BAGSTER'S GESENIUS. BAGSTER'S COMPLETE EDITION OF GESENIUS HEBREW AND CHALDEE LEXICON. In large, clear, and perfect type. Translated and edited with additions and corrections, by S. P. Tregelles, LL.D.

In this edition great care has been taken to guard the student from Neologian tendencies by suitable remarks whenever needed.

"The careful revial to which the Lexicon has been subjected by a faithful and Orthodox translator exceedingly enhances the practical value of this edition."
—*Edinburgh Ecclesiastical Journal.*

Small 4to, half bound.............................$7 00

*BAGSTER'S. ANALYTICAL HEBREW AND CHALDEE LEXICON. With an Alphabetical Arrangement of every Word in Old Testament, &c., &c. By B. Davidson. 1 vol. small 4to, half-bound $11.00

NEW POCKET HEBREW AND ENGLISH LEXICON The arrangement of this Manual Lexicon combines two things—the etymological order of roots and the alphabetical order of words. This arrangement tends to lead the learner onward; for, as he becomes more at home with roots and derivatives, he learns to turn at once to the root, without first searching for the particular word in its alphabetic order. 1 vol. 18mo, cloth...............................$2 00

"This is the most beautiful, and at the same time the most correct and perfect Manual Hebrew Lexicon we have ever used."—*Eclectic Review.*

IRON, METALLURGY, &c.

BODEMANN. A TREATISE ON THE ASSAYING OF LEAD, SILVER, COPPER, GOLD, AND MERCURY. By Bodemann & Kerl. Translated by W. A. Goodyear. 1 vol. 12mo, $2 50

CROOKES. A PRACTICAL TREATISE ON METALLURGY. Adapted from the last German edition of Prof. Kerl's Metallurgy. By William Crookes and Ernst Rohrig. In three vols. thick 8vo. Price.....................................$30 00

Separately. Vol. 1. Lead, Silver, Zinc, Cadmium, Tin, Mercury, Bismuth, Antimony, Nickel, Arsenic, Gold, Platinum, and Sulphur.................................$10 00
Vol. 2. Copper and Iron...........................10 00
Vol. 3. Steel, Fuel, and Supplement...............10 00

DUNLAP. WILEY'S AMERICAN IRON TRADE MANUAL of the leading Iron Industries of the United States. With a description of the Blast Furnaces, Rolling Mills, Bessemer Steel Works, Crucible Steel Works, Car Wheel and Car Works, Locomotive Works, Steam Engine and Machine Works, Iron Bridge Works, Stove Foundries, &c., giving their location and capacity of product. With some account of Iron Ores. By Thomas Dunlap, of Philadelphia. 1 vol. 4to. Price to subscribers $7 50

FAIRBAIRN CAST AND WROUGHT IRON FOR BUILDING. By Wm. Fairbairn. 8vo, cloth......................$2 00

FRENCH. HISTORY OF IRON TRADE, FROM 1621 TO 1857. By B. F. French. 8vo, cloth.........................$2 00

KIRKWOOD COLLECTION OF REPORTS (CONDENSED) AND OPINIONS OF CHEMISTS IN REGARD TO THE USE OF LEAD PIPE FOR SERVICE PIPE, in the Distribution of Water for the Supply of Cities. By I. P. Kirkwood, C.E. 8vo, cloth......................$1 50

MACHINISTS—MECHANICS.

FITZGERALD THE BOSTON MACHINIST. A complete School for the Apprentice and Advanced Machinist. By W. Fitzgerald. 1 vol. 18mo, cloth...............................$0 75

HOLLY. SAW FILING. The Art of Saw Filing Scientifically Treated and Explained. With Directions for putting in order all kinds of Saws, from a Jeweller's Saw to a Steam Saw-mill. Illustrated by forty-four engravings. Third edition. By H. W. Holly. 1 vol. 18mo, cloth.........................$0 75

TURNINC, &c. **LATHE, THE, AND ITS USES, ETC.; or, Instruction in the Art of Turning Wood and Metal.** Including a description of the most modern appliances for the ornamentation of plane and curved surfaces, with a description also of an entirely novel form of *Lathe* for Eccentric and Rose Engine Turning, a Lathe and Turning Machine combined, and other valuable matter relating to the Art. 1 vol. 8vo, copiously illustrated.   Including Supplement.   8vo, cloth......$7 00

"The most complete work on the subject ever published."—*American Artisan.*
"Here is an invaluable book to the practical workman and amateur."—*London Weekly Times.*

TURNINC, &c. **SUPPLEMENT AND INDEX TO LATHE AND ITS USES.** Large type. Paper, 8vo..................$0 90

WILLIS. **PRINCIPLES OF MECHANISM.** Designed for the use of Students in the Universities and for Engineering Students generally. By Robert Willis, M.D., F.R.S., President of the British Association for the Advancement of Science, &c., &c. Second edition, enlarged. 1 vol. 8vo, cloth.........$7 50

₊ It ought to be in every large Machine Workshop Office, in every School of Mechanical Engineering at least, and in the hands of every Professor of Mechanics, &c.—Prof. S. EDWARD WARREN.

MANUFACTURES.

BOOTH. **NEW AND COMPLETE CLOCK AND WATCH MAKERS' MANUAL.** Comprising descriptions of the various gearings, escapements, and Compensations now in use in French, Swiss, and English clocks and watches, Patents, Tools, etc., with directions for cleaning and repairing. With numerous engravings. Compiled from the French, with an Appendix containing a History of Clock and Watch Making in America. By Mary L. Booth. With numerous plates. 1 vol. 12mo, cloth..................... ....$2 00

CELDARD. **HANDBOOK ON COTTON MANUFACTURE; or, A Guide to Machine-Building, Spinning, and Weaving.** With practical examples, all needful calculations, and many useful and important tables. The whole intended to be a complete yet compact authority for the manufacture of cotton. By James Geldard. With steel engravings. 1 vol. 12mo, cloth...................................$2 50

MEDICAL, &c.

BULL. **HINTS TO MOTHERS FOR THE MANAGEMENT OF HEALTH DURING THE PERIOD OF PREG-NANCY, AND IN THE LYING-IN ROOM.** With an exposure of popular errors in connection with those subjects. By Thomas Bull, M.D. 1 vol. 12mo, cloth..........$1 00

FRANCKE **OUTLINES OF A NEW THEORY OF DISEASE,** applied to Hydropathy, showing that water is the only true remedy. With observations on the errors committed in the practice of Hydropathy, notes on the cure of cholera by cold water, and a critique on Priessnitz's mode of treatment. Intended for popular use. By the late H. Francke. Translated from the German by Robert Blakie, M.D. 1 vol. 12mo, cloth...$1 50

CREEN **A TREATISE ON DISEASES OF THE AIR PASSAGES.** Comprising an inquiry into the History, Pathology, Causes. and Treatment of those Affections of the Throat called Bronchitis, Chronic Laryngitis, Clergyman's Sore Throat, etc., etc. By Horace Green, M.D. Fourth edition, revised and enlarged. 1 vol. 8vo, cloth....................................$3 00

" **A PRACTICAL TREATISE ON PULMONARY TUBER-CULOSIS,** embracing its History, Pathology, and Treatment. By Horace Green, M.D. Colored plates. 1 vol. 8vo, cloth................................$5 00

GREEN.   **OBSERVATIONS ON THE PATHOLOGY OF CROUP**
With Remarks on its Treatment by Topical Medications. By
Horace Green, M.D. 1 vol. 8vo, cloth..............$1 25

"   **ON THE SURGICAL TREATMENT OF POLYPI OF
THE LARYNX, AND ŒDEMA OF THE GLOTTIS.**
By Horace Green, M.D. 1 vol. 8vo.................$1 25

"   **FAVORITE PRESCRIPTIONS OF LIVING PRACTI
TIONERS.** With a Toxicological Table, exhibiting the
Symptoms of Poisoning, the Antidotes for each Poison, and
the Test proper for their detection. By Horace Green. 1
vol. 8vo, cloth............................... . ..$2 50

TILT.   **ON THE PRESERVATION OF THE HEALTH OF
WOMEN AT THE CRITICAL PERIODS OF LIFE.**
By E. G. Tilt, M.D. 1 vol. 18mo, cloth.............$0 50

VON DUBEN.   **GUSTAF VON DUBEN'S TREATISE ON MICRO-
SCOPICAL DIAGNOSIS.** With 71 engravings. Translated, with additions, by Prof. Louis Bauer, M.D. 1 vol. 8vo.
cloth...$1 00

MINERALOGY.

BRUSH.   **MANUAL OF DETERMINATIVE MINERALOGY,** with
an Introduction on Blow-Pipe Analysis. By Prof. Geo. J.
Brush. 1 vol. 8vo.................................$3 00

DANA.   **DESCRIPTIVE MINERALOGY.** Comprising the most recent Discoveries. Fifth edition. Almost entirely re-written
and greatly enlarged. Containing nearly 900 pages 8vo, and
upwards of 600 wood engravings. By Prof. J. Dana.
Cloth............................... ....$10 00
"We have used a good many works on Mineralogy, but have met with none that
begin to compare with this in fulness of plan, detail, and execution."—
American Journal of Mining.

DANA & BRUSH. **APPENDIXES TO DANA'S MINERALOGY,** bringing
the work down to 1875. 8vo... ....$1 00

DANA.   **DETERMINATIVE MINERALOGY.** (See Brush's Blow-
Pipe, etc.). 1 vol. 8vo, cloth.....................$3 00

"   **A TEXT-BOOK OF MINERALOGY.** 1 vol. (In prepa
ration.)

MISCELLANEOUS.

BAILEY.   **THE NEW TALE OF A TUB.** An adventure in verse. By
F. W. N. Bailey. With illustrations. 1 vol. 8vo......$0 75

CARLYLE.   **ON HEROES, HERO-WORSHIP, AND THE HEROIC IN
HISTORY.** Six Lectures. Reported, with emendations and
additions. By Thomas Carlyle. 1 vol. 12mo, cloth...$0 75

CATLIN.   **THE BREATH OF LIFE;** or, Mal-Respiration and its
Effects upon the Enjoyments and Life of Man. By Geo.
Catlin. With numerous wood engravings. 1 vol. 8vo, $0 75

CHEEVER.   **CAPITAL PUNISHMENT.** A Defence of. By Rev. George
B. Cheever, D.D. Cloth..........................$0 50

"   **HILL DIFFICULTY,** and other Miscellanies. By Rev.
George B. Cheever, D.D. 1 vol. 12mo, cloth........$1 00

"   **JOURNAL OF THE PILGRIMS AT PLYMOUTH ROCK.**
By Geo. B. Cheever, D.D. 1 vol. 12mo, cloth.......$1 00

,   **WANDERINGS OF A PILGRIM IN THE ALPS.** By
George B. Cheever, D.D. 1 vol. 12mo, cloth........$1 00

"   **WANDERINGS OF THE RIVER OF THE WATER OF
LIFE.** By Rev. Dr. George B. Cheever. 1 vol. 12mo,
cloth $1 00

CONYBEARE.   **ON INFIDELITY.** 12mo, cloth......................1 00

CHILD'S BOOK **OF FAVORITE STORIES.** Large colored plates. 4to,
cloth................. ....$1 50

EDWARDS   **FREE TOWN LIBRARIES.** The Formation, Management and History in Britain, France, Germany, and America. Together with brief notices of book-collectors, and of the respective places of deposit of their surviving collections. By Edward Edwards. 1 vol. thick 8vo.............$4 00

GREEN.   **THE PENTATEUCH VINDICATED FROM THE AS·PERSIONS OF BISHOP COLENSO.** By Wm. Henry Green, Prof. Theological Seminary, Princeton, N. J. 1 vol. 12mo, cloth..$1 25

GOURAUD.   **PHRENO-MNEMOTECHNY;** or, The Art of Memory. The series of Lectures explanatory of the principles of the system. By Francis Fauvel-Gouraud. 1 vol. 8vo, cloth, $2 00

"   **PHRENO-MNEMOTECHNIC DICTIONARY.** Being a Philosophical Classification of all the Homophonic Words of the English Language. To be used in the application of the Phreno-Mnemotechnic Principles. By Francis Fauvel-Gouraud. 1 vol. 8vo, cloth...........................$2 00

HEIGHWAY   **LEILA ADA.** 12mo, cloth......................... .. **1 00**

"   **LEILA ADA'S RELATIVES.** 12mo, cloth.......... **1 00**

KELLY.   **CATALOGUE OF AMERICAN BOOKS.** The American Catalogue of Books, from January, 1861, to January, 1866. Compiled by James Kelly. 1 vol. 8vo, net cash.......$5 00

"   **CATALOGUE OF AMERICAN BOOKS.** The American Catalogue of Books from January, 1866, to January, 1871. Compiled by James Kelly. 1 vol. 8vo, net...........$7 50

MAVER S   **COLLECTION OF GENUINE SCOTTISH MELODIES.** For the Piano-Forte or Harmonium, in keys suitable for the voice. Harmonized by C. H. Morine. Edited by Geo. Alexander. 1 vol. 4to, half calf......................$10 00

PARKER.   **QUADRATURE OF THE CIRCLE.** Containing demonstrations of the errors of Geometers in finding the Approximations in Use; and including Lectures on Polar Magnetism and Non-Existence of Projectile Forces in Nature. By John A. Parker. 1 vol. 8vo, cloth.................$2 50

STORY OF   **A POCKET BIBLE.** Illustrated. 12mo, cloth.......$1 00

SVEDELIUS.   **HAND-BOOK FOR CHARCOAL BURNERS.** Translated from the Swedish by Prof. R. B. Anderson, and edited by Prof. W. J. L. Nicodemus, C.E. 1 vol., 12mo. Plates. Cloth..$1 50

TUPPER.   **PROVERBIAL PHILOSOPHY.** 12mo..............$1 00

WALTON   **THE COMPLETE ANGLER;** or, The Contemplative Man's
& COTTON.   Recreation, by Isaac Walton, and Instructions how to Angle for a Trout or Grayling in a Clear Stream, by Charles Cotton, with copious notes, for the most part original. A bibliographical preface, giving an account of fishing and Fishing Books, from the earliest antiquity to the time of Walton, and a notice of Cotton and his writings, by Rev. Dr. Bethune. To which is added an appendix, including the most complete catalogue of books in angling ever printed, &c. Also a general index to the whole work. 1 vol. 12mo, cloth..$3 00

WILLIAMS.   **THE MIDDLE KINGDOM.** A Survey of the Geography, Government, Education, Social Life, Arts, Religion, &c., of the Chinese Empire and its Inhabitants. With a new map of the Empire. By S. Wells Williams. Fourth edition, in 2 vols...$4 00

RUSKIN'S WORKS.

Uniform in size and style.

RUSKIN **MODERN PAINTERS.** 5 vols. tinted paper, bevelled boards, plates, in box.....................................$18 00

" **MODERN PAINTERS.** 5 vols. half calf............ 27 00

" " " " without plates........ 12 00

" " " " half calf, 20 00

Vol. 1.—Part 1. General Principles. Part 2. Truth.

Vol. 2.—Part 3. Of Ideas of Beauty.

Vol. 3.—Part 4. Of Many Things.

Vol. 4.—Part 5. Of Mountain Beauty.

Vol. 5.—Part 6. Leaf Beauty. Part 7. Of Cloud Beauty. Part 8. Ideas of Relation of Invention. Formal. Part 9. Ideas of Relation of Invention, Spiritual.

" **STONES OF VENICE.** 3 vols., on tinted paper, bevelled boards, in box....................................$7 00

" **STONES OF VENICE.** 3 vols., on tinted paper, half calf..$12 00

" **STONES OF VENICE.** 3 vols., cloth.............. 7 00

Vol. 1.—The Foundations.

Vol. 2.—The Sea Stories.

Vol. 3.—The Fall.

" **SEVEN LAMPS OF ARCHITECTURE.** With illustrations, drawn and etched by the authors. 1 vol. 12mo, cloth, $1 75

" **LECTURES ON ARCHITECTURE AND PAINTING.** With illustrations drawn by the author. 1 vol. 12mo, cloth....................................$1 50

" **THE TWO PATHS.** Being Lectures on Art, and its Application to Decoration and Manufacture. With plates and cuts. 1 vol. 12mo, cloth....................$1 25

" **THE ELEMENTS OF DRAWING.** In Three Letters to Beginners. With illustrations drawn by the author. 1 vol. 12mo, cloth..$1 00

" **THE ELEMENTS OF PERSPECTIVE.** Arranged for the use of Schools. 1 vol. 12mo, cloth..................$1 00

" **THE POLITICAL ECONOMY OF ART.** 1 vol. 12mo, cloth...$1 00

" **PRE-RAPHAELITISM.**

NOTES ON THE CONSTRUCTION OF SHEEPFOLDS. } 1 vol. 12mo, cloth, $1 00

KING OF THE GOLDEN RIVER; or, The Black Brothers. A Legend of Stiria.

RUSKIN **SESAME AND LILIES.** Three Lectures on Books, Women, &c. 1. Of Kings' Treasuries. 2. Of Queens' Gardens. 3. Of the Mystery of Life. 1 vol. 12mo, cloth.........$1 50

" **AN INQUIRY INTO SOME OF THE CONDITIONS AT PRESENT AFFECTING "THE STUDY OF ARCHITECTURE" IN OUR SCHOOLS.** 1 vol. 12mo, paper...................................$0 15

" **THE ETHICS OF THE DUST.** Ten Lectures to Little Housewives, on the Elements of Crystallization. 1 vol. 12mo, cloth....................................$1 25

" **"UNTO THIS LAST."** Four Essays on the First Principles of Political Economy. 1 vol. 12mo, cloth...........$1 00

RUSKIN

THE CROWN OF WILD OLIVE. Three Lectures on Work, Traffic, and War. 1 vol. 12mo. cloth.................$1 00

"

TIME AND TIDE BY WEARE AND TYNE. Twenty-five Letters to a Workingman on the Laws of Work. 1 vol. 12mo, cloth..... ....$1 00

"

THE QUEEN OF THE AIR. Being a Study of the Greek Myths of Cloud and Storm. 1 vol. 12mo, cloth $1 0C

ʻ

LECTURES ON ART. 1 vol. 12mo, cloth............ 1 00

＊

FORS CLAVIGERA. Letters to the Workmen and Labourers of Great Britain. Part 1. 1 vol. 12mo, cloth, plates, $1 00

"

FORS CLAVIGERA. Letters to the Workmen and Labourers of Great Britain. Part 2. 1 vol. 12mo, cloth, plates, $1 00

MUNERA PULVERIS. Six Essays on the Elements of Political Economy. ˙1 vol. 12mo, cloth...... ....$1 00

"

ARATRA PENTELICI. Six Lectures on the Elements of Sculpture, given before the University of Oxford. By John Ruskin. 12mo, cloth, $1 50, or with plates. $3 00

"

THE EAGLE'S NEST. Ten Lectures on the relation of Natural Science to Art. 1 vol. 12mo...............$1 50

"

THE POETRY OF ARCHITECTURE: Villa and Cottage. With numerous plates. By Kata Phusin. 1 vol. 12mo, cloth.................. ....$1 50

Kata Phusin is the supposed Nom de Plume of John Ruskin.

"

FORS CLAVIGERA. Letters to the Workmen and Laborers of great Britain. Part 3. 1 vol. 12mo, cloth.......

"

LOVES MEINE. Lectures on Greek and English Birds. By John Ruskin. Plates, cloth......................$0 75

"

ARIADNE FLORENTINA. Lectures on Wood and Metal Engraving. By John Ruskin. Cloth..............$1 00

"

FRONDES AGRESTES, AND MORNINGS IN FLORENCE. 12mo, cloth............................$1 00

"

THE TRUE AND THE BEAUTIFUL IN NATURE, ART, MORALS, AND RELIGION. Selected from the Works of John Ruskin, A.M. With a notice of the author by Mrs. L. C. Tuthill. Portrait. 1 vol. 12mo, cloth, plain, $2.00; cloth extra, gilt head........ ....$2 5C

"

ART CULTURE. Consisting of the Laws of Art selected from the Works of John Ruskin, and compiled by Rev. W. H Platt. A beautiful volume. with many illustrations. 1 vol. 12mo, cloth, extra gilt head....................$3 00

"

Do. Do. School edition. 1 vol. 12mo, plates, cloth..$2 50

"

PRECIOUS THOUGHTS: Moral and Religious. Gathered from the Works of John Ruskin, A.M. By Mrs. L. C. Tuthill. 1 vol. 12mo, cloth, plain, $1.50. Extra cloth, gilt head..$2 00

"

SELECTIONS FROM THE WRITINGS OF JOHN RUSKIN. 1 vol. 12mo, cloth, plain, $2.00. Extra cloth, gilt head..$2 5C

"

SESAME AND LILIES. 1 vol. 12mo, ex. cloth, gilt head,$1 75

"

ETHICS OF THE DUST. 12mo, extra cloth, gilt head, 1 75

"

CROWN OF WILD OLIVE. 12mo, extra cloth, gilt head, 1 50

RUSKIN'S BEAUTIES.

"

THE TRUE AND BEAUTIFUL ⎱

PRECIOUS THOUGHTS. ⎰ 3 vols., in box, cloth extra, gilt head.........$6 00

CHOICE SELECTIONS. do., half calf...10 0C

RUSKIN'S POPULAR VOLUMES.

RUSKIN

CROWN OF WILD OLIVE. SESAME AND LILIES. QUEEN OF THE AIR. ETHICS OF THE DUST. 4 vols. In box, cloth extra, gilt head.................$6 00

RUSKIN'S WORKS.
Revised edition.

" Vol. 1.—**SESAME AND LILIES.** Three Lectures. By John Ruskin, LL.D. 1. Of King's Treasuries. 2. Of Queens' Gardens. 3. Of the Mystery of Life. 1 vol. 8vo, cloth, $2.00. Large paper.............................$2 50

" Vol. 2.—**MUNERA PULVERIS.** Six Essays on the Elements of Political Economy. By John Ruskin. 1 volume 8vo, cloth. $2 00
Large paper.. 2 50

" Vol. 3.—**ARATRA PENTELICI.** Six Lectures on the Ele ments of Sculpture, given before the University of Oxford. By John Ruskin. 1 vol. 8vo.......................$4 00
Large paper...................................... 4 50

RUSKIN—COMPLETE WORKS.

THE COMPLETE WORKS OF JOHN RUSKIN. 20 vols., extra cloth, in a box.·.$44 00
Ditto   30 vols., extra cloth. *Plates*... 52 00
Ditto   Bound in 20 vols., half calf.   do.... 78 00

SHIP-BUILDING, &c.

BOURNE.

A TREATISE ON THE SCREW PROPELLER, SCREW VESSELS, AND SCREW ENGINES, as adapted for Purposes of Peace and War. Illustrated by numerous wood-cuts and engravings. By John Bourne. New edition. 1867. 1 vol. 4to, cloth, $18.00; half russia................$24 00

WATTS.

RANKINE (W. J. M.) AND OTHERS. Ship-Building, Theo retical and Practical, consisting of the Hydraulics of Ship-Building, or Buoyancy, Stability, Speed and Design—The Geometry of Ship-Building, or Modelling, Drawing, and Laying Off—Strength of Materials as applied to Ship-Building —Practical Ship-Building—Masts, Sails, and Rigging—Marine Steam Engineering—Ship-Building for Purposes of War. By Isaac Watts. C.B., W. J. M. Rankine, C.B,, Frederick K. Barnes. James Robert Napier, etc. Illustrated with numerous fine engravings and woodcuts. Complete in 30 numbers, boards, $35.00; 1 vol. folio, cloth, $37.50; half russia, $40 00

WILSON (T. D.) SHIP-BUILDING, THEORETICAL AND PRACTICAL. In Five Divisions.—Division I. Naval Architecture. II. Lay ing Down and Taking off Ships. III. Ship-Building. IV. Masts and Spar Making. V. Vocabulary of Terms used— intended as a Text-Book and for Practical Use in Public and Private Ship-Yards. By Theo. D. Wilson, Assistant Naval Constructor, U. S. Navy; Instructor of Naval Construction, U. S. Naval Academy; Member of the Institution of Naval Architects, England. With numerous plates, lithographic and wood. 1 vol. 8vo. $7 50

SOAP.

MORFIT.

A PRACTICAL TREATISE ON THE MANUFACTURE OF SOAPS. With numerous wood-cuts and elaborate work ing drawings. By Campbell Morfit, M.D., F.C.S. 1 vol. 8vo...$20 00

STEAM ENGINE.

TROWBRIDGE.

TABLES, WITH EXPLANATIONS, OF THE NON CONDENSING STATIONARY STEAM ENGINE, and of High-Pressure Steam Boilers. By Prof. W. P. Trow bridge, of Yale College Scientific School. 1 vol. 4to. plates...$2 50

" **HEAT AS A SOURCE OF POWER:** with applications of general principles to the construction of Steam Generators. An introduction to the study of Heat Engines. By W. P. Trowbridge. Prof. Sheffield Scientific School, Yale College. Profusely illustrated. 1 vol. 8vo. cloth......... ...$3 50

TEXT-BOOKS for Use of U. S. Naval Academy.

COOKE.   **A TEXT-BOOK OF NAVAL ORDNANCE AND GUN-NERY.** Prepared for the use of the Cadet Midshipmen at the United States Naval Academy. By A. P. Cooke, Com. U. S. N. One thick volume, illustrated by about 400 fine cuts. Cloth...................................$12.50

RICE & JOHNSON.   **ELEMENTS OF THE DIFFERENTIAL CALCULUS,** founded on the Method of Rates or Fluxions. 8vo.

WILSON.   **SHIP-BUILDING, THEORETICAL AND PRACTICAL.** By T. D. Wilson. (See page 15.) 8vo, cloth........$7.50

TURNING, &c.

THE LATHE,   **AND ITS USES, ETC.** On Instructions in the Art of Turning Wood and Metal. Including a description of the most modern appliances for the ornamentation of plane and curved surfaces. With a description, also, of an entirely novel form of Lathe for Eccentric and Rose Engine Turning, a Lathe and Turning Machine combined, and other valuable matter relating to the Art. 1 vol. 8vo, copiously illustrated, cloth.........$7 00

"   **SUPPLEMENT AND INDEX TO SAME.** Paper...$0 90

VENTILATION.

LEEDS (L. W.).   **A TREATISE ON VENTILATION.** Comprising Seven Lectures delivered before the Franklin Institute, showing the great want of improved methods of Ventilation in our buildings, giving the chemical and physiological process of respiration, comparing the effects of the various methods of heating and lighting upon the ventilation, &c. Illustrated by many plans of all classes of public and private buildings, showing their present defects, and the best means of improving them. By Lewis W. Leeds. 1 vol. 8vo, with numerous wood-cuts and colored plates. Cloth........$2 50

"It ought to be in the hands of every family in the country."—*Technologist.*

"Nothing could be clearer than the author's exposition of the principles of the principles and practice of both good and bad ventilation."—*Van Nostrand's Engineering Magazine.*

"The work is every way worthy of the widest circulation."—*Scientific American.*

REID.   **VENTILATION IN AMERICAN DWELLINGS.** With a series of diagrams presenting examples in different classes of habitations. By David Boswell Reid, M.D. To which is added an introductory outline of the progress of improvement in ventilation. By Elisha Harris, M.D. 1 vol. 12mo, $1 50

J. W. & SONS are Agents for and keep in stock

SAMUEL BAGSTER & SONS' PUBLICATIONS,

LONDON TRACT SOCIETY PUBLICATIONS,

COLLINS' SONS & CO.'S BIBLES,

MURRAY'S TRAVELLER'S GUIDES,

WEALE'S SCIENTIFIC SERIES

Full Catalogues gratis on application.

J. W. & SONS import to order, for the TRADE AND PUBLIC.

BOOKS, PERIODICALS, &c.,

FROM

ENGLAND, FRANCE, AND GERMANY.

*** JOHN WILEY & SONS' Complete Classified Catalogues of the most valuable and latest scientific publications, Parts I. and II., 8vo, mailed to order on the receipt of 10 cts.

www.ingramcontent.com/pod-product-compliance
Lightning Source LLC
Chambersburg PA
CBHW021947220326
41599CB00012BA/1357